Antler Art for Baskets and Gourds

Betsey Sloan

Schiffer Publishing Ltd
4880 Lower Valley Road, Atglen, Pennsylvania 19310

Dedication

To my past, present, and future students...
May we continue to learn, create, and laugh together.

Acknowledgments

I thank all the artists who submitted their work as projects and for the Gallery.

I also thank Eidnes Furs, who provided me with several antlers for the antler identification section, and Jennifer VanAllen, a truly gifted and talented photographer.

Last but not least, thank you to the folks at Schiffer Publishing for having faith in me as a writer and allowing me to choose the topics of my books.

Front cover main art courtesy of Jill Choate.

Other Schiffer Books by the Author:
InLace Techniques: Resin Inlay for Gourd and Wood Crafts,978-0-7643-3330-9, $12.99

Other Schiffer Books on Related Subjects:
Basket Inspiration for Makers & Collectors, 978-0-7643-3468-9, $34.99
Let's Weave Color into Baskets, 0-88740-056-6, $12.95
Gourd Crafts: 6 Projects & Patterns, 978-0-7643-2825-1, $14.95

Library of Congress Control Number: 2010938963

Designed by Mark David Bowyer
Type set in Humanist521 BT

ISBN: 978-0-7643-3615-7
Printed in China

Schiffer Books are available at special discounts for bulk purchases for sales promotions or premiums. Special editions, including personalized covers, corporate imprints, and excerpts can be created in large quantities for special needs. For more information contact the publisher:

Published by Schiffer Publishing Ltd.
4880 Lower Valley Road
Atglen, PA 19310
Phone: (610) 593-1777; Fax: (610) 593-2002
E-mail: Info@schifferbooks.com

For the largest selection of fine reference books on this and related subjects, please visit our web site at:
www.schifferbooks.com
We are always looking for people to write books on new and related subjects. If you have an idea for a book please contact us at the above address.

This book may be purchased from the publisher.
Include $5.00 for shipping.
Please try your bookstore first.
You may write for a free catalog.

In Europe, Schiffer books are distributed by
Bushwood Books
6 Marksbury Ave.
Kew Gardens
Surrey TW9 4JF England
Phone: 44 (0) 20 8392 8585; Fax: 44 (0) 20 8392 9876
E-mail: info@bushwoodbooks.co.uk
Website: www.bushwoodbooks.co.uk

Contents

Introduction

History of Antlers

Antler vs. Horn

When I first started working with antlers, I often referred to the antler as a "horn." While doing this in class, some of the other students would quietly correct me by saying "antler." I gradually understood why. Here is the reason: an antler is dropped, or shed, each year from the male animal. Each spring, an antler is grown to replace the shed. A horn is not dropped or shed — ever. Cows have horns; deer have antlers. Sheep have horns; elk have racks or antlers.

Moose, caribou, and fallow deer have antlers that are palmated, or flat, in certain areas, like a hand's palm with the fingers spread out. "Points" on an antler refer to the number of tips or outgrowths. Thus, a four-pointer would be antlers that have two outgrowths on each antler. When a hunter refers to a deer as being a "spike horn" — a deer with only two long antlers with no other projections from them — technically it should be "spike antlers," but we won't argue the point. Some artists also refer to these points as prongs or tines.

A "natural shed" refers to an antler that has been shed or dropped by the male animal usually in late winter. The end of the antler on a shed is bumpy where it detached from the head. A "cut" antler refers to an antler that was cut off the male's head. The end of the antler, or butt end, is smooth and sometimes shows these cut marks. This may mean the animal is no longer with us or that the antler came from a deer farm where they are sometimes removed so the animals do not endanger each other during mating season.

A cut antler is shown on the left. A natural shed antler, with its nubby end, is on the right.

White Tails, Mule Ears, and the Big Guys

Antlers and the animals that wear them got their start about twenty-five million years ago, near the beginning of the time period of great mammal development. Science terms this period the Miocene. Much of the earth at this time was tropical and semi-tropical. With all this humidity and warmth, plant growth was rampant. The early generations of even-toed hoofed animals, although antlerless, navigated through this dense jungle.

There is much speculation as to why and how these antlerless animals first began to grow antlers. Was it to differentiate the sexes? Was it for defense against other animals? Was the growth a result of head butting or was it genetic programming? One thing is certain, we will never know for sure.

Since man first became a meat eater, antlers and hides of deer and other animals have been of major importance to the progression of mankind. Animals offered not only food, but also skins for warmth, fat for protection, slivers of bone for sewing, and long bones for weapons.

Whitetails and Mule Deer

The deer genus comprises two species in North America: the whitetails and mule deer. The Columbian blacktail of the Northern Pacific and British Columbia areas is a slightly smaller variation of the mule deer. A third member of the mule deer family is the Sitka blacktail. Today, there are close to thirty sub-species of whitetail and mule presently living above the Mexican border.

Through the centuries, all the species of deer — as well as moose, elk, and caribou — have ranged over the North American continent. Some are thought to have migrated from Europe and Africa before the continents drifted apart.

The whitetails' most striking feature, other than the heavy antlers of the mature male or buck, is the foot-long tail from which it gets its name. While brown on the top or outer side, it is snowy white underneath, but when alarmed and the tail is turned upright, the rump and tail look like a "flag."

In contrast, the muley's tail is narrow and black towards the end of its tip. The main key difference with the muley is its huge, mulish ears that constantly twitch while gathering the faintest of sounds from far distances.

Another way the muley and the whitetail differ is the style of locomotion. The whitetail runs by pushing off alternately its front and rear legs in long, graceful bounds. By contrast, the muley prefers to launch itself with all four legs at once, similar to a pogo stick and reminiscent of several types of African antelopes.

Whitetail antlers have points that grow from only one side of the curve and generally have a more pronounced curve to them than mule antlers.

Mule deer antlers can have multiple points growing from any side of the curve and are typically more "Y" shaped than whitetail antlers.

The defining character of the whitetail antler is that the points, tines, or prongs sprout individually from the main beam, just as in the elk. The mature male whitetail antler, therefore, is curved in one direction with the points growing off of one side only. The muley buck, however, has a bifurcated, main beam forking into two smaller beams; in the mature bucks, these smaller beams can be divided again, giving the mule antler a distinctive "Y" formation.

Other Antlered Creatures

We have yet to discuss the elusive elk, moose, reindeer, or caribou. Imagine meeting a moose standing some seven feet high at the shoulders and more than nine feet in length. Then imagine their antlers! In a ratio of whitetails or muleys to these larger creatures, their numbers are much smaller; as a result, their antlers are harder to locate and cost more to purchase. In addition, to create a basket on one of these wonderful antlers is a true challenge. Not only does it take experience, but a *lot* of time and material. The effort, though, is worth it if you can get your hands on some of these larger antlers.

Lastly, there are deer that are farm-raised, although some of their species also remain in the wild. These include fallow, red, and Sitka. These farm-raised deer are used for meat or are sold to hunting preserves. They are also raised for their antlers.

Presently, in addition to the demand for antlers used for artwork, lamps, and other decorative items, there is a huge market for antlers in velvet. Velvet refers to the state where antlers have begun to grow, but have not yet reached the full year's growth. The antler at this stage is covered in a soft velvet-like substance, usually rubbed off by the male animal once the antler has reached full maturity. The male rubs off the itchy velvet against trees or anything handy at his antler height, thus marking not only his territory, but making tree scrapings that allow hunters to know that a male deer is somewhere in the area. Antlers still in velvet are prized by Asian cultures as a component in ancient medicines. Deer that are farm raised may have their antlers cut off while in velvet. Others may be illegally taken since most hunting seasons are well after the velvet stage.

Red deer, associated with the elk family, have straight or slightly curved antlers. They might also have multiple points at the head joint.

Fallow deer antlers have one or two palmated sections and can run from 15 to 30-plus inches in length.

Moose antlers are extremely large and heavy compared to their smaller cousins, with a huge palmated section from which multiple points can sprout.

Elk antlers are graceful, with a sweeping swing from back to front. The one shown here is being compared to a very curved whitetail antler.

Caribou, or reindeer, antlers have a broad, palmated area near their tips as well as a second set of shorter palmated tines that grow forward from the head.

State Laws and You

Good News and Bad News

I was hoping to include the state laws as they relate to the use of whitetail antlers. I say whitetail specifically as this is the only deer that is prevalent in most of the United States and, therefore, native or indigenous. Antlers from farm-raised deer and or deer not native to your state usually can be used without any restrictions.

However, after receiving responses from the first few states I contacted, I realized that I could never publish a listing that would be accurate. The guidelines on antler use are constantly changing. I feared that by the time this book got to print the information would be outdated and misleading to my readers. For instance, the Midwest is currently in flux with the laws as they relate to antler use and even moving deer meat or deer parts across state lines. There is a chance of spreading a disease that has become prevalent among the whitetails, similar to Mad Cow, although not quite the same, and not passed to humans.

Most artisans get their antlers from many different places. Friends, family, reputable antler suppliers, junk stores, and garage sales are all excellent sources for both cut and naturally shed antlers. However, before using antlers in your basketry and gourd art for classes or for sale, check with your state government about laws regarding their use (the Wildlife Division, Division of Conservation, Fish and Game, or whatever department that handles rules on using whitetail antlers). If you're going to use any part of an antler in your work, it is *your* responsibility to research this. In most cases, personal use is okay, although the antler still must be legally obtained.

Here is what could happen if you are in a state that prohibits whitetail usage: Let's say, for example, you've used several whitetail antlers in your work and you are at a very crowded fine art/craft show. A state official, from let's assume the Wildlife Division, was there checking out the show and saw the antlers. This official could close the entire show because you have broken the law by using an antler illegally. This almost happened a number of years ago at a basket conference where one of the instructors innocently used whitetail antlers for a basket class — without realizing the law in that particular state forbade their usage. The conference came close to being closed down until a compromise was reached. Can't you just picture eight hundred irate basket weavers if the conference had been closed down?

The good news, though, is that I have brought this matter to your attention, so now you can legally pursue adding antlers to your basketry and gourd art. The bad news is that you will have to research your state's website to discover the proper government agency that regulates antler usage, make a few phone calls, and, in the end, you might be disappointed.

Section One

Tools and Techniques

Chapter One

Antlers in Basketry

Drilling Guidelines for Basketry

While compiling the projects for this book, I discovered other ways to drill and frame for baskets. Some folks drill the antler so the center spoke comes out of the butt end of the antler rather than from underneath the antler. Some artists bring this center spoke around and drill another hole at one of the points and glue in the center spoke. Others don't drill all the way through the butt end and simply glue the rims in place.

I wasn't taught these methods, so what follows are *my* own personal styles of drilling and framing. If you have been taught differently or use another method, that is fine. I've always believed that there is more than one way to reach a conclusion. As long as we all get to the same place in the end, it doesn't matter how we've arrived.

There are a few firm rules, however, that I adhere to and which I pass on to you for whatever they are worth.

Rule One: Never drill too close to the antler tip, as it will surely break.

Rule Two: Avoid using a "soft" material (yarn, seagrass, leather, etc.) to hold the ends of the spokes in place. Over time they will work their way out. The only exception to this may be a wall basket; as a purely visual piece of art, no strain will be placed on the basket.

Rule Three: Each basket you make will look different from the one before.

Rule Four: Always make sure the antler is securely attached to the basket or gourd.

Rule Five: Wear eye protection and a respiratory mask when drilling.

Drilling antlers, whether to accept reed for basket framing or to make holes for attaching the antler to a gourd, is an art form. The most important thing is the angle of the drill hole, especially for baskets. This knowledge will come after you have worked with antlers for awhile.

If you're just starting to use antlers in your art, here's what you need to know: first, the antlers smell when they are drilled. The heat of the drill bit warms up the bone, giving off a rather smelly smoke. Secondly, the small shards of bone that are created by drilling can be very sharp, so take care and be sure to wear protective gear.

Drill Bits

It would be great if we could pick any old antler and drill a hole straight through it for a basket rim, but this isn't the case. You have to visualize where the reed will flow, from the butt end (the part that is attached to the head) around to and through the points or tips, and then back to the butt end. If the angle is too sharp and the reed becomes stressed, it might snap or split. The same is true for drilling through the points or tips. If the angle is not correct, the reed will have a problem feeding through it and, again, may snap over time.

The size of the drill bit can vary depending on the size of the reed being used as the rim. Basket makers understand that #6 round reed can often be slightly larger or smaller in diameter than the previous pound of #6 they used. Likewise, a #5 round reed can be closer to a #6, so you have to choose a bit that fits the diameter of the reed you have in your hand, not necessarily what is shown in the chart below.

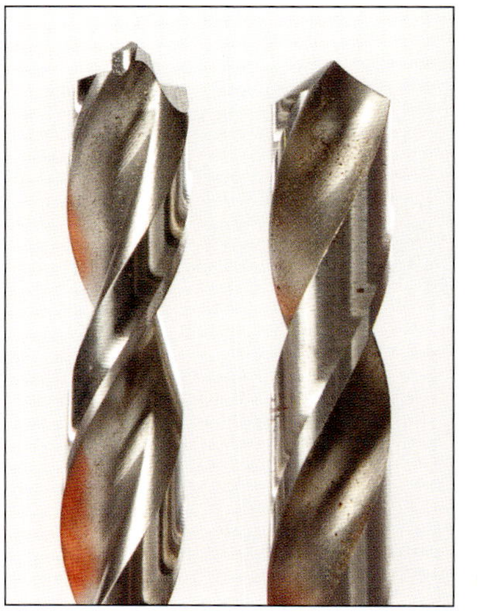

Bits for drilling antlers should have a pointed end so that they grasp the bone. The tungsten bit on the left has a good sharp tip. The bit on the right is a regular drill bit and may slip on the antler.

The type of drill bit is important, too. If you can locate Black and Decker Bullet® bits, these are excellent. They have a little prong on the drill bit end that helps bite into the antler. Titanium drill bits also have this prong, but they can be more expensive. Regular drill bits without this prong can slip off the antler and make drilling more difficult.

Here are the typical bit sizes for smoked or natural round reed. Be willing to adjust the bit size depending on the reed to be used.

Reed Size	Bit Size
#2 and #3 round reed	1/8th
#4 round reed	9/64th
#5 round reed	5/32nd
#6 round reed	3/16th
#7 round reed	7/32nd
#8 round reed	17/64th
Larger round reed	check against larger bits

Drilling Instructions

The following directions are for a standard antler basket with the antler used as a handle. However, the descriptions of how to determine angles and how to drill can apply to any style of basket calling for the antler to be woven upon as the handle.

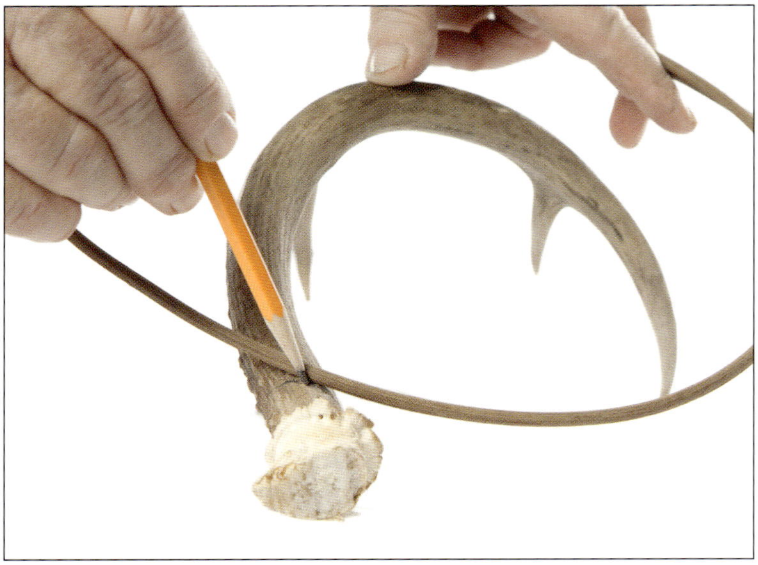

Use a length of rim material to determine where the reed will "flow" around the antler, from butt to tip(s). Mark the angle with pencil.

What I do first when preparing an antler for a basket is to picture how the basket will be woven on that particular antler. What is the curve like? Can I drill through the points at a sufficient angle to accept the rims? How large can the basket opening be? Once this is in my mind, I place the antler on a cutting board. I have one that is well used and it acts as a buffer between the drill bit and the table.

- First, I use a length of rim material and eyeball where I think the reed will flow, from the butt end to the points and back.
- Then I mark with a pencil the angle where the drilling needs to be.
- Finally, turn the antler on its side on your cutting board and carefully start the drill bit so that it digs into the antler. Once the bit has started to score the antler, start drilling along the predetermined line.

Using the pencil marks as a guide, drill through the butt end of the antler.

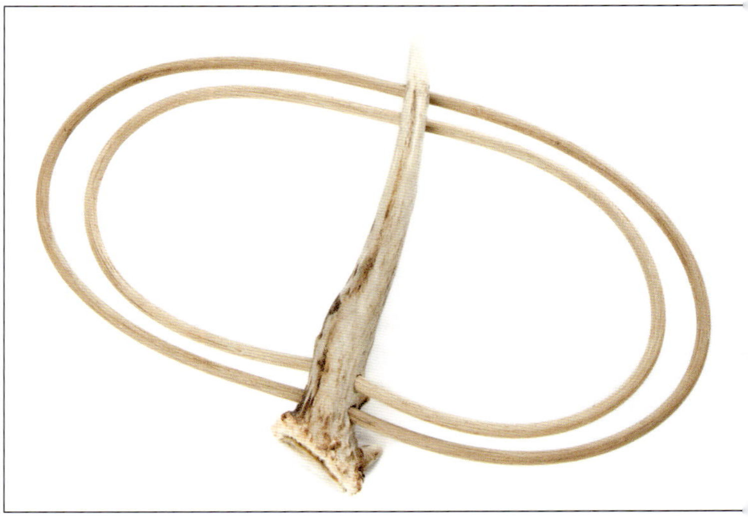

The two rims on this small antler show the correct frame spacing.

You have to be prepared for several things. The antler consists of many layers, including the outer bone and an inner soft core. You can feel the drill bit reaching that center core because the bit will "give" a little. The bit will then grab the opposite bone side and continue through. When you reach that second side and go all the way through the antler, you will feel a "pop" as you clear that second side with no more resistance. However, do NOT shut off the drill. Simply back out the drill bit with the drill still running.

Note: If you stop the drill once you feel the bit clear the opposite side, chances are you will end up with a lovely antler with a drill bit stuck in its butt end. Reversing the drill may work or you can try using a pair of pliers to reverse the bit out of the antler. What happens is that the heat of drilling expands the bone and once the heat is stopped, the bone cools, but stays in the expanded shape, thus grabbing the drill bit.

Drill a second hole about 1/2" from the first, to hold the second rim. I usually weave with two upper rims whenever possible as I think it makes a stronger basket. For extra large antlers, I may even use three or four rims.

Once you have drilled the butt end, insert a length of rim reed through the one of the holes and bring the reed around to the point end of the antler to determine the angle of the drilling through these points. I always like to secure the basket to at least two of the points, if possible. If the antler has only one point or has a point that cannot be utilized, then go with only one point.

Determine the angle that the reed needs to pass through the points, one to the other, and then back to the butt end. This is often the trickiest part of framing an antler basket. If the angle between the points is off, the reed will have a difficult time running through the points, so consider the angles carefully. Mark with a pencil and drill. You will need two holes if you are using two rims. If you don't have the space for two holes, drill one for the inner rim and let the second rim flow underneath the antler tips. The weaving will hold it in place.

Most drilling on the point ends should be well above the tips. Just exactly where will be determined by the size of the reed being used as well as where the reed will flow from one point to the other. Sounds vague, I know, but once you've done a few antlers, this will become second nature to you.

Now for the bottom center spoke. Flip the antler over and pencil mark a spot underneath where the outermost hole was drilled at the butt end. Drill a hole at about a 45-degree angle, deep enough to hold the spoke in place. This angle will allow the center spoke to "pouch" out at the butt end, creating a pleasing shape to the basket. Your drilling may go through into the previous drilled holes, and this is okay. If you drilled a hole straight into the antler, without this angle, then the center spoke has a tendency of collapsing inwards at the butt end.

Drill the center spoke hole in the butt end of the antler. Angle the drill so that the spoke can gently curve outwards from the antler to create a good shape for the weaving.

Framing a Typical Antler Basket

Finally, once all the drilling is completed, frame the basket. Run the reed from the butt end, through the points, and back to the butt. The outside hole closest to the butt end corresponds to the holes towards the very tip of the antler. Likewise, the inner rim is the second hole in from the butt end, corresponding to the second holes up from the tip end. Keep these two rims the same distance apart all the way around. Once you have determined the shape, trim the reed and glue at the butt end. I use super glue with an activator because I want to make sure the bond is secure and will not be affected by water or by pulling on it while weaving. Again, this framing will come with practice. If you get the opening too large, the sides will droop down, similar to a hearth basket. If they are too small, the basket will look out of proportion to the antler.

Glue a length of reed in the center spoke hole and bring the reed around to the point ends. Bring the length of reed around to the point end and judge where this center spoke will end up in the basket (the depth of the basket as well as the shape). I usually select one point that will be the "front" of the basket, as it is esthetically more pleasing than the butt end, and where I might add embellishments like beads or pods. Tie the center spoke underneath the selected point with waxed linen securely, but not so tight that you can't adjust it if necessary while weaving. Trim the center spoke so that you have two or three inches to play with. The end of this spoke, as well as the waxed linen, can be removed once the basket is woven.

This fully framed antler basket has two rims and a center spoke.

This close-up of one of the antler tips shows how the center spoke is tied off underneath the spoke. Some basket weavers prefer to drill another hole and glue the spoke in place, but I prefer tying it off so I can adjust it if necessary while weaving.

Drilling for Woven Bowls

Using an antler for a basket base is a great way to create a unique basket. For a typical bowl style basket, the antler selected should be able to lay flat on the table, upside-down; in other words, with the points heading upwards. You can also weave a basket just the opposite of this, with the points facing down towards the table and the entire basket woven upwards. For demonstration purposes, the first position was selected.

- Mark with pencil the path on the antler for a series of holes to accept the spoke material — the line should go from the butt end, along the top ridge of the antler (outside curve), down the antler as the line approaches the point end just before the antler begins to branch out, and then along the inside curve just above where the antler hits the table.
- Select a drill bit that will fit the reed used as spokes. I typically use a #3 or #4 smoked round reed as the spoke material. Whatever reed you choose, be sure it is pliable enough to bend into the modified Japanese border once you get to the upper rim of the basket.

The angle for drilling the spokes is important, as this will determine how the basket will rise from the antler. To allow the weaving to build upwards along the curve of the points, the holes for the spokes should be angled correctly to allow the reed to bend naturally in that direction.

There also needs to be a bottom to the basket (located in the curve of the antler), so the drill holes for the spokes should allow the spokes to lay flat on the table at the start of the weaving. They can be curved upwards at some point with the tension placed on the weaving. See the project titled "Antler Bowl" in Chapter Eight for more information on framing and weaving.

To create a bottom for the antler bowl, drill holes in the side of the antler at the curve so that the spokes can lie flat on the table to create a base. After the base is woven, these spokes are gently turned upwards.

The angle of the four spokes at the tips and butt end on an antler bowl is critical. The spokes need to lie flat against the antler tips until the basket is ready to be brought upwards. Likewise, the butt end spokes need to allow the weaving to pouch out, creating a pleasing bowl shape.

This close-up of an antler bowl shows the spoke work and weaving as it comes from the antler, positioned with the points downward. *From Nancy Jones.*

Drilling an Antler Wall Basket

There are various ways to attach an antler to a wall basket. One is to weave the basket and then drill and attach the antler with leather or waxed linen wherever the antler will fit on the basket. The way I am showing here is how I was taught:

- Using a very curved whitetail antler, drill two holes: one at the butt end and one on the outside of the curved point, at an angle to create the lower rib of the basket.
- Glue a length of #6 reed that will create a pleasing basket in proportion to the antler. The length of the reed will depend on the size of the antler and your "eye" as to the balance between the basket and antler.

For additional information on how to create the upper rim (called a "D" ring) and weave with both the antler and rim, see "Antler Wall Pouch" in Chapter Seven.

The initial spoke for an antler wall basket is typically drilled underneath the curve with one center spoke glued in place, creating the lowest point for the basket weaving.

Chapter Two

Using Spokes and Weavers

The Beginning Weave

Now that you have your standard antler basket framed and are ready to start weaving, soak some long lengths of #1, #2 or #3 round reed; they can be dyed, smoked, or natural. The size will depend on the overall size of the basket you're weaving. You will need to weave on the rims and center/bottom spoke in order to create "pockets" into which the new spokes will be added.

For an antler basket with two upper rims (referred to as inside and outside rims) and a center spoke, here is the process for the beginning weave. I usually start at the butt end, but you can start at the point end if you wish.

- Insert one end of the soaked weaver underneath the center spoke. Hold the end in place with your left hand (if you are right-handed). Take the weaver and bring it up to the right and go over the outside rim, then under the inside rim, around through the center of the basket, bringing the reed now over the top of the inside rim. You have just reversed direction.
- Now, go underneath the outside rim, back to the center spoke where you will go over, and then to the left side of the basket.
- Bring the reed under the left side outside rim, over the inside rim, and around and underneath the inside rim to reverse direction.
- Go over the outside rim and bring the reed back to the center where the weaving will be under. The first row of weaving should now stay in place without you holding it.
- This completes one continuous row of over/under weaving. Repeat for about 5 to 6 rows until you have spaces to insert new spokes.

A close-up of the first few rows of weaving shows the half-moon design created by the over/under continuous weave. Notice the weaving is over the center spoke and rims, not the antler.

The tension on the weaver for the beginning weave is important. It cannot be so strong that it misshapes the basket and it cannot be so loose that there is a gap between the rim and the weaver. Sometimes it may be necessary to turn over the innermost rim twice before heading back to the opposite side. You want to create a half-moon weave as well as good spaces in which to add spokes. How many double turns will depend on how you drilled the antler and the distance between the center spoke and the rims. If after about four rows the weaving resembles a "U" instead of the half moon, weave a few more rows and execute double turns on the inside rim.

This style of basket weaving, or continuous weaving, can be a strain on your shoulders because you have to carefully pull the entire reed through as you execute each step, over and under. Rest as needed. Keep the weavers damp, but not overly wet. When using dyed reed, it's best to just dampen the reed and then remove it from the water pail. Never leave reed soaking in the water unless it's a large sized reed (#4, #5) that needs to be bent double for other styles of basketry.

Once you have woven about 1 to 1-1/2 inches on one side of the basket, let the weaver hang (do not cut it off) and go to the other side of the antler and repeat the same pattern of weaving. When done, you will add four spokes (see "Adding Spokes" in next section) into the spaces created by the weaving.

Adding Spokes

Spokes refer to the lengths of reed that are inserted in the side and bottom of typical antler baskets. The ends of the spokes are tucked into the weaving, which create "V" shaped areas. Spoke ends should be trimmed down before they are inserted into the weaving. The reed can be cut with scissors, on an angle. It can also be ground in a pencil sharpener or trimmed with a reed rasp. Whatever method you use, be sure that the end of the added spoke "snicks" into place. Don't you love basket terms?

You should be able to hear the new spoke squeak into place up against the old spoke. This means it is tight into the basket and the end is securely pushed against the old spoke so that the new end won't show. I prefer cutting the round reed with scissors, on an angle. The cut side is the side that is slipped up against the old spoke. I usually do one side of the spoke and then feed the reed across to the opposite side; measure, cut, and inset.

You can only add one spoke at a time into a space. Add spokes as they are needed and when the weaving allows it. Also remember that, depending on the curve of the antler, one side of the basket may require more spokes than the other. No two baskets (and antlers) are alike.

To hold new spokes in position, re-wet the weaver and continue where you left off. To add new weavers, backtrack for one spoke, ending the old weaver and starting the new weaver to the inside of the basket. These will be trimmed once all of the weaving is done. Weave one side of the basket with a material and then mirror that material on the other side of the basket, working towards the center bottom. When adding spokes, eyeball each one and push it into position where it will most likely end up with more weaving completed. Does the spoke seem too short or too long? If so, remove it and insert another one.

It's important to understand that once you have added a spoke into the basket, even if the same number are added on each side of the basket, the over-under weaving will be interrupted. This means that the weaving will mirror the previous row of weaving, or you will see two over-weaves and two under-weaves on the same spoke. If you can possibly add two spokes at a time in the same general location, then the weaving will be "off" only between those two new added spokes. The best time to add a single spoke to a side is on the upward turn of weaving. On the bottom area, always try to add two at a time, one either side of the center, so the weaving follows the correct pattern at the sides. Once you begin to work with spokes, you will grasp this concept.

If you find a gap at the bottom of the basket while weaving the final material, simply insert a spoke as best as you can. The rule of thumb for spacing between spokes at the bottom center of an antler basket is no more than two finger widths between spokes. If a spoke is too long at the very bottom and will not allow the basket to sit properly, cut the offending spoke in half on a sharp angle, overlap the two pieces, and weave together.

Likewise, if you find one of the spokes you inserted is too short and you've woven it in place already, you can change it for a new spoke. Whatever you do, don't pull on the spoke and slip it out of its holding space. Your weaving will then collapse and it's almost impossible to repair this type of error. To change a spoke that is too short after it has been woven in place, insert a packing tool or screwdriver into the weaving where the offending spoke is attached. Carefully pull the spoke out. Insert a newly trimmed piece of reed. Be sure the new spoke is securely pushed into the space. Remove the sharp tool. Go to the opposite side of the basket. Gauge the length of the new spoke and cut. Using the same sharp tool, insert it into the old spoke's weaving, remove the old spoke, and insert the new spoke.

Whatever type of antler basket you are creating, be sure to pay attention to the basket as you weave and add spokes. Constantly check to be sure your weaving and the added spokes are creating the shape you desire. With traditional antler baskets, don't let the butt end collapse inwards. The pouching out of the butt end will help stabilize the basket once completed.

Spokes should be sharpened prior to inserting them into the weaving. Shown here is one of the first spokes is being inserted. Notice that the cut side of the new spoke is being inserted against the old spoke.

Here is a typical antler basket about two-thirds woven. See the pattern created by the weaving and how the spokes have been added as weaving has allowed.

If you find that the basket won't sit flat on the table when it's done, soak the bottom and sides (depending on the materials used. If the material is all natural, water won't hurt it). Then carefully push down on the antler from above and "help" the spokes flatten out. This can be dangerous and you might not want to try it — a spoke can snap if it is not wet enough — but if it's sufficiently wet, the bottom will flatten. Further, if you put a weight in it while it is drying, it will retain that flat bottom. It's best, however, to check your spoke additions each time so that you don't have to use this procedure!

Packing the Weave

Many of the projects in this book refer to "packing" the sides or bottoms of antler baskets that use spokes to create the shape. If you can visualize a framed antler basket as it is woven, the basket needs to extend beyond the upper rims, not only for a pleasing design, but also to allow the basket to sit well. This means that as you continuously weave back and forth, from side to side, the rims will be used up or woven upon faster than the very bottom of the basket. You could continue to weave, using all the rim space, but this creates an inverted "V" shaped bottom that may or may not be pleasing. I was taught to get my sides of the basket as straight as possible across the bottom, leaving about three to four inches to be filled with palm inflorescence for strength. The amount of space would depend on the size of the basket, of course.

In order to create even weaving on the sides, it's necessary to turn on the spokes below the rim. Where and how often will depend on the overall shape of the basket. For instance, my typical basket would weave as follows ("rows" means from rim to opposite rim):

- Smoked #2 reed, six rows
- Dyed #3 reed, six rows
- Seagrass, six rows

Now switch to vine rattan and begin packing by NOT going to the upper rim after one complete rim to rim weave, but turn on the spoke below it. Weave back to the other side of the basket, and again turn one below the rim.

On the next row of weaving, go back over to the other side, but turn on the spoke below where you previously turned. Repeat to the other side and keep doing this, dropping one spoke from each side as you weave until you can't turn anymore at the very bottom of the basket.

Now, simply reverse the process, weaving back up a spoke as you go back and forth. The last row would go from rim to rim, making the vine rattan part of the overall reed that shows at the rim level. This type of packing ends up creating a fairly straight swath at the basket bottom. At this time, I weave straight across this swath, from one side to the other, until all the space is filled in.

Packing is basically a feel you will develop as you weave antler baskets. Turning on various spokes while weaving will become second nature. This is a personal preference, of course. You may not mind the inverted "V" that is created by not packing as you weave each side of the basket. However, you will still have to do some packing, or turning on spokes other than the rim. The only difference is that one is done at the sides of the basket where it is less noticeable, and the other at the very center of the sides, where it will show. Try both methods and see which is more pleasing to your eye, and which is easier to weave.

TIP: Flat reed, especially anything larger than 3/64th, does not pack as well as round reed or round vine or seagrass because it tends to bunch up, one row over the other, as you try forcing it to pack. In addition, there is a visible space where you have made selective turns. Keep this in mind when selecting materials with which to weave.

Bottom packing is clearly visible in this antler basket.

Here is a good example of packing. You can see that the rim was only used for about seven or eight turns while the spokes below the rim were used until all the space was filled with weaving.

On an antler pouch, you can see how packing of the weavers was used to fill in the space.

Marianne Barnes used the packing technique in her antler basket, both on the sides and at the base.

Chapter Three

Attaching Antlers

Basket Attachments

There are ways to attach antlers to baskets other than drilling for handles or basket bases. Some antlers are attached after the basket is woven or they are attached into the basket weaving, as the basket is started, up near the rim. Here are some examples of how different artists have attached antlers to their baskets:

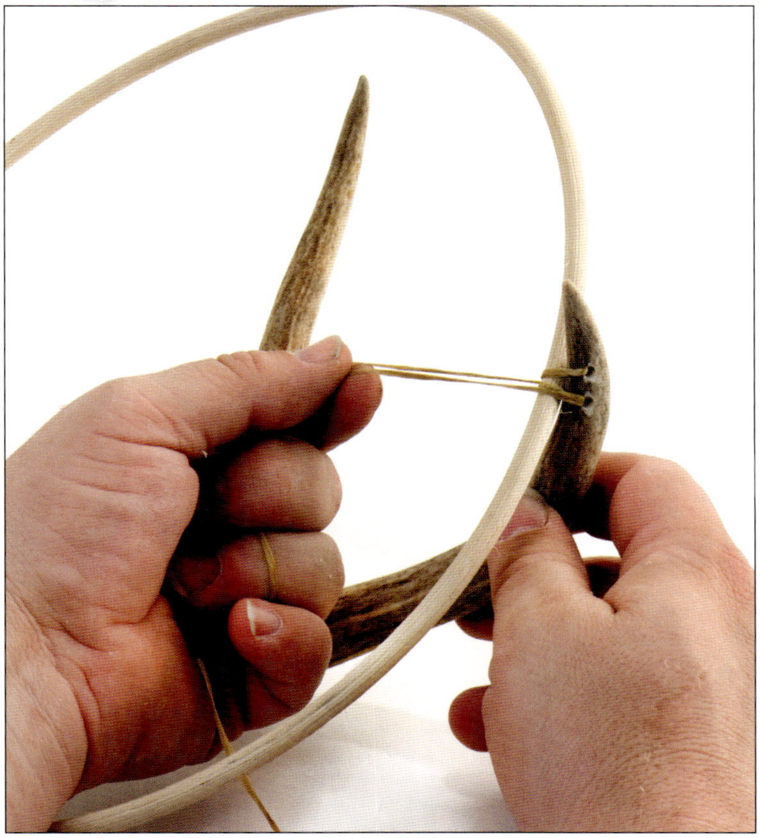

While framing his antler wall basket, Bob Hosea was unable to drill a hole at the antler tip large enough for the rim material. Instead, he drilled a small hole and tied the rim to the antler. *Photography and copyright by Brittany Hosea*.

Nancy Jones attached her antler to the basket by lashing it to the center spokes of her egg-shaped basket.

Here are three examples of attaching antlers to a woven basket. Joyce Storto used flat reed to attach the antler securely at three separate places to the basket rim.

Karen Zane used round reed to more firmly attach this antler to her basket even though the antler was incorporated with the basket rim to start the basket.

At the base of this antler bowl, Arlene Zaboznik chose to weave over the antler in two areas.

Gourd Attachments

As with antler baskets and rims, the angle of the drill hole is crucial in attaching a gourd to an antler. If the gourd is expected to sit on top of the antler as a base for the artwork and dowels are used to attach the two, the angle from the antler through the gourd must be precise or the gourd will not sit well. Less critical are the holes for attaching antlers to gourds at the sides or top, although most artists would not want these holes to show and will cover them with lacing and other embellishments.

The most important thing about using antlers with gourds is finding the right antler to fit the gourd. One that is too long or too heavy will cause the gourd to fall over. One that is too small will look out of proportion. Just because you like a certain antler and a certain gourd does not mean the two will be a good "fit." An antler that slides all over the place will continue to do so, even after drilling and tying it in place. I swear there is still some spirit alive in the antler that says, "nope," but in any event, the antler should be securely fastened to the gourd. Some artists use glue and a small trough cut in the gourd rim; some use leather or attach the antler into coiling.

I prefer to make small drill holes in the antler and use waxed linen or similar material to tie the antler to the gourd, either directly or to the coiling or rim material. Sometimes I'll use leather strips and lash the antler in place, but most likely I would have drilled and tied the antler first with waxed linen just to make it secure. Each antler is unique unto itself — and the same can be said for gourds. If the antler is to sit on top of the gourd, it has to rest there, preferably without any help. If it slips or causes the gourd to tip over, it is the wrong antler for that gourd.

Here are some examples of how different artists attach their antlers to gourd work:

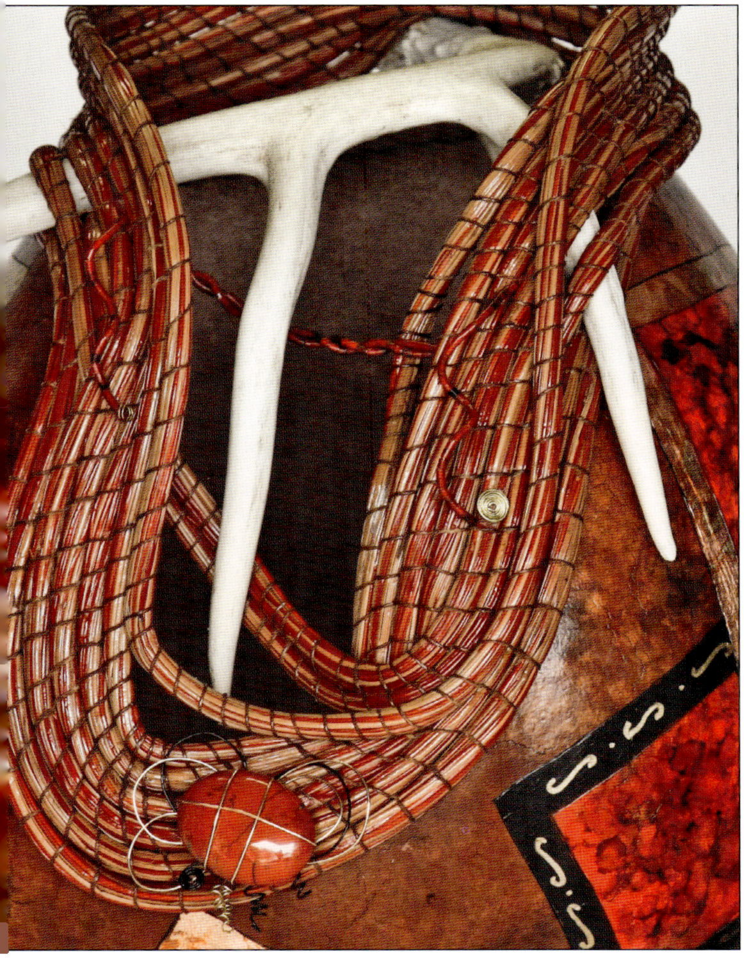

Coiling on this gourd shows the intricate interweave of the antler into the material. It is securely held by the coiling going over and under several areas of the antler. *From Patty Sorenson; photography and copyright by Bill Clements.*

Another way to attach antlers to gourds is by using bolts or screws. Jill Nelson chose to use two bolts on the inside of her gourd mask to hold the antlers in position.

Holes were drilled into the gourd and leather lashing was used to tie the antler to the rim as a means of securing the antler to the gourd in this project.

After drilling small holes into the antler, the gourd can be utilized to make a secure attachment. Jaynie Barnes has chosen this method to attach an antler to her finished gourd.

An antler can be curved over the gourd rim, or in this case, secured right through the gourd itself. Here, the artist drilled through both sides of a gourd opening, allowing the antler to fit snugly into position. *From Pam Montgomery*.

Antler Gourd Bowl

For the gourd bowl used with the fallow antler tip in Chapter Eight, here is the procedure I used to put the pieces together:

- Put the raw gourd bowl on the antler exactly where you want it to sit. Make sure the gourd "fits" the antler, and that there are two or three points where the gourd hits the antler at the bottom or even side.
- Carefully drill through the gourd, from the inside, until it just hits the antler so as to mark it. Use a bit that is the same size as the dowel being used.
- Remove the gourd, and drill into the antler for about 1/4", or deep enough to allow gluing of the dowel piece. After temporarily inserting a length of dowel, about 2 inches worth, into the new hole, through the gourd and into the antler, repeat the process in another spot so that the gourd is held tightly at least in two locations.
- Connect the gourd to the antler with the pieces of dowel. Mark on the dowels where they need to be cut, so they are just above the gourd bottom on the inside. Remove the gourd, remove the dowels, and trim the dowels.
- Now proceed with the artwork, knowing that the gourd will fit where you want it. If you created the artwork first and then drilled holes, there is a chance of making mistakes and the gourd might be ruined.
- Once the artwork is done and the finished applied to the gourd, glue the dowels to the antler and to the gourd bottom.
- Use some type of material to cover the dowels at the gourd bottom. This could be leather, paper lining, clay, or a good strong paint that will cover both the wood and the gourd interior.

TIP: For the last step listed above, I've found dyes and inks don't work well, as the wood takes color differently than the gourd lining. A circle of leather, covering the entire gourd bottom, works well.

Chapter Four

Weaving Materials

Soft vs. Hard

Whether using reed, vines, philodendron sheaths, cane, seagrass, or other materials primarily for basketry where spokes are to be added, it is important to understand the texture or strength of each element used. "Soft" materials are those that are usually woven to provide texture or color. These include seagrass, yarns, philodendron sheaths or other leaf fibers, and coconut coir. "Hard" materials mean any of the round reeds, whether dyed, smoked, or natural, as well as palm inflorescence, honeysuckle, or other strong vines like rattan that help shape and hold the spokes in place.

Flat reeds may act as either soft or hard, depending on their size. The smaller the flat reed, say 11/64th or 3/16th, the tighter the weave; therefore, spokes can be added and held in place. The larger flat reeds, like the 1/4 or 1/2" sizes, the more "wiggle room" the spoke will have and eventually the spoke could work its way out of the weaving. This is why I suggest always weaving the first 12 rows or so of an antler basket with a hard material so that the spokes are held firmly in place. Of course, the larger the basket, the more "hard" material you will need to use for strength and shaping ability, as well as to hide the spoke ends.

This soft vs. hard is not so critical in gourd work since most of the weaving is secondary to the already firm gourd. Keep in mind, however, that when coiling and working the antler into the coils, the coiling must be strong enough to keep the antler in place while not endangering the shape of the coiling. Weight is an issue, too. Make sure the antler is inserted into the coiling in a way that it doesn't tip the gourd over. So many things to consider!

Hard materials for weaving include round reed, some small gauge flat reeds, palm inflorescence, and other firm vines that will hold spokes in position during the weaving.

Soft materials are those materials that might eventually allow the spokes to work their way through the weaving and eventually out of the basket. These include yarns, seagrass, and large gauge flat reeds.

I often will do some basket weaving on the top of a gourd bowl. I like to use yarns — the more funky the better — for my first few rows of twining. The yarn helps cover the holes where the reed comes out of the gourd and also adds some fun element to the piece. I don't expect the yarn, however, to help shape the upper basket, so I switch to a #2 or #3 round reed for that and then drill and tie on the antler at two points as an embellishment.

Gathering Materials

Much enjoyment can be gained from gathering your own materials to use in both basketry and gourd art, but it can also be a lot of hard work. While in Florida, during the early years of my retirement in the mid 1990s, a friend and I used to go around the various towns on "green dump" day — the day when cuttings from landscaping work was left by the curb to be picked up and hauled to the green dump. We would leave early in the morning and roam through the neighborhoods, stopping to pick up the long Queen Palm stalks. We would eventually cut off the long stems, which are then called palm inflorescence. Homeowners who spotted us would usually come out and say we were welcome to their trash, but what were we going to do with it? We began to keep a sample of our antler baskets with us, showing the bottoms where we had woven the palm for strength.

Once we returned home with our stash, we would divvy up the stalks and begin the hard work of cutting each stem off them and then bundling the stems together to dry. Each bundle is kept separate because each palm stalk produces stems that can vary in size, color, and length. Palm inflorescence is wonderful for antler basket bottoms. The stems are soaked in water for several hours or overnight until pliable and then woven. The stems are a little rough on the hands and sometimes the very fat ends need to be trimmed because they won't bend easily even after hours of soaking.

We also used to visit the local green dump where all the rest of the trimmings were taken. My friend and I would sneak in and gather cuttings from coconut tree trimmings. We would later break off the coco bloom part, heat them in the oven for about a half hour on very low temperatures, and then store them for using as embellishments, but my husband said he wouldn't bail me out if we were caught — we weren't exactly trespassing — so we didn't often break the law! (And I do recommend following "No Trespassing" signs for your own safety and well-being.)

Honeysuckle vines and other vines like Kudzu can all be used in basketry. Harvesting them can be time consuming, though, as you need good length and clean runners for weaving. There are also a number of cultivated and wild leaves that can be used for weaving, such as cattail and iris. If this interests you, go to the local library or online and research when's the best time to pick these vines and leaves and how to dry them.

Reeds, rattan, seagrass, and other materials are as near as your local basket supplier. Some of the material I use is mentioned in the "Suppliers' Directory" section of this book. These are considered natural materials. Reed comes from the bamboo that grows at the rate of about 12 to 15" a day in Indonesia. It is a renewable resource. Likewise, the seagrass is actually a sea grass that has been coiled into a fine rope. Vine rattan is the vine that holds the bamboo to the trees as it grows. Cane is the very outside of the bamboo stalk; reed comes from the inside.

Always use as fresh a coil of reed as possible. Reed that has aged over a few years or more is often brittle and no amount of soaking in water will soften it back up.

Note about soaking reeds: Whether you are soaking natural or dyed reeds in the water bucket, keep pets away from the water. The chemicals used in processing reed are harmful to small animals.

Section Two

The Projects

Chapter Five

Baskets

Small Antler Basket

Susan McCluskey Ashley

Materials

- Small antler, approximately 7" with "Y" prong
- #5 and #2 round smoked or stained reed
- 3/64th flat reed
- 1/4" flat smoked or stained reed
- 1/8" suede lacing or other weaving material

Drilling

Using the butt or shed end of the antler and a length of #5 round reed, determine where to drill for the rims. I usually use two pieces of reed, thereby creating two rims: one close to the shed end; the other on the same level, but towards the forked end. With a pencil, mark the entrance and exit for the drill holes once you have visualized where the basket curve will look best on the antler. Place the first mark as close to the shed end as possible and the second about 1/4" away, towards the point end.

Carefully drill the holes completely through the antler using the pencil lines as a guide. Use a drill bit that matches the size of the #5 reed. You now have holes for both an outside and inside rim.

Now for the tricky part: Eyeball the angle needed to bring the rim around and "through" the part of the antler about 1" above the forked part. Mark these spots and the angle carefully. Lay the antler on a table so that you can brace the drill bit on the antler and drill about a 1/2" deep hole. Do NOT drill all the way through. Now, turn the antler over and mark the opposing spots. Again, drill, but not all the way through. Make sure the angle you are drilling will allow the reed to create a pleasing yet sturdy rim.

Finally, drill three holes underneath the antler, at the shed end, to hold the bottom spokes. Turn the antler upside down and mark three spots on this underside. The marks should be on the same line as the outside rim hole, only underneath the antler. Drill three holes to fit the #5 reed. The holes should be at a slight angle so that the spoke is allowed to pouch outwards from the end of the antler once the weaving has started. You might drill into that outside spoke hole, but this is not a problem.

Creating the Rims and Bottom Spoke

Pull the two rims through the holes in the butt end and feed them around to the holes in the point end. Adjust the rims until they are evenly spaced apart. I use either super glue or a glue gun to hold the spokes in place. The basket shown here has one rim side longer than the other, but that was more pleasing to me, given the curve of the antler.

For the bottom three spokes, glue in three lengths of #5 reed. In the basket shown, the center spoke was 11". The halfway spoke to the longer side of the basket was 12-1/2" with the other halfway spoke only 10". The length will depend on your "eye" and how you picture the basket woven on the antler.

Weaving the Basket

Start at either the shed end or the forked end and, using a damp length of #2 round reed, weave two to three rows of over/under continuous weave. Let the weaver hang in place and go to the opposite end and, with a fresh length of #2, weave two or three rows. As with any egg style basket, insert spoke as soon as possible. I sharpened mine with a pencil sharpener, adding an even number each time. Add spokes and weave at the shed end. Bring the new spoke over, insert it into the fork end, and weave. Continue adding and weaving, changing materials as desired. Always check how the basket sits after you have added spokes. The antler should remain level and the basket should grow out and under, creating a platform on which to sit.

Weave the remainder of the basket as you would any egg basket. I like to use the small material at first, where the spokes are closer together, and end with a wider, flat reed when the spokes get further apart. Varying the different materials as you go creates an interesting look. Use soft material, like suede lacing, for accent, but do not use it in an area that will have to hold the spokes in place.

This shot of Susan's basket shows her precise weaving and spoke additions at the butt end of the antler.

Ribbed Basket with Antler Handle

Marianne Barnes

Because all antlers differ greatly, every antler basket will be different. You will need to take your round reed and plan the shape of the frame by running one end of the reed from the butt end of the antler to wherever you determine they will cross over the point end. Visualize how the basket will be shaped on the antler and proceed.

Materials

- Shed antler with two tips/points
- #4 or #5 round reed
- 1/4" flat natural reed
- 1/4" leather strips
- Coconut hair rope (coir)
- Small palm inflorescence (stalks)
- Philodendron sheath

Preparing the Frame for Basket

Select a drill bit to fit the #4 or #5 reed and drill a hole on each side of the butt end of the antler, but do not go all the way through. Feed in one end of a length of reed and see where the reed will flow over the points. Mark the areas. Drill a hole completely through the tip ends, but not too close to the tips, as they break easily. I use only one rim with my small to medium baskets.

Next, drill two holes on the underneath side of the antler: one will be at the butt end, where you drill the first hole, only underneath; the second, underneath the longest antler tip.

Glue one end of the round reed in one of the butt end holes, then take the reed around to the first tip and through the hole. Feed the reed through the second tip and back to the butt end. Determine the shape of the rim and then trim and glue the reed at the butt end. Take another length of reed and determine how tall you want the basket to be. It should be in relation to the overall size of the basket; not too tall, but not too shallow, either. When you have determined this depth, trim the reed and glue into place. This is now your center spoke.

Weaving the Basket

The first thing I do is to weave a simple half-God's Eye at each end of the frame. To do this, take one piece of 1/4" flat reed and taper the end for about 2". Put the tapered end under the center spoke at either end of the antler, then take the weaver to the top right rim, and wrap the flat reed over the top and back to the center spoke. Go over the center spoke. Then go to the left round reed rim; go under the left rim, around the top of the rim, and back down to the center again. Go under the center and repeat the over/under for about five rows. Leave this weaver hanging and go to the opposite end of the antler and, with a fresh weaver, repeat the process.

Marianne decided to add some spokes between the antler points once the basket was woven. She made sure to weave into the existing rim and carried through the color/texture theme so it looked like a continuation of the basket.

Using scissors, scraper, or sharpener, trim the ends of some round reed and insert them in the spaces created by the weaving. Bring the new spoke over to the opposite end of the antler and determine the length, trim, and insert. The first two spokes added at the bottom will be the same length as the center spoke. Add an even number of spokes on each side. Continue the half-God's Eye pattern without going over or under these new spokes. Simply hold them in place. After about 15 to 16 rows, end the weaving by tapering the flat reed and slipping it up into the half-God's Eye. Do this at both ends of the antler.

Now you will start weaving rows, using the added spokes as well as new ones along the way. Taper the end of another length of flat reed. Start somewhere along the side of the basket and start an over/under weave pattern. Weave from one side of the basket to the other for about four to five rows. Repeat on the other end of the basket. Add more spokes as needed, weaving them over/under as the pattern calls for. Make sure you leave enough room to weave when adding spokes and use long enough weavers to give the basket strength.

Change materials as desired, but only add soft materials where the spoke ends may not poke out. I used Coconut coir or rope, then back to some flat reed, and then several rows of leather strips. The palm inflorescence makes a good bottom material since it is very strong when dry. Be sure to wet it with hot water to make it more pliable.

As you weave back and forth, you will usually create a "V" shape on one side of the basket. You will find you can't go to the upper rim of the basket anymore. Simply turn on the spoke below the rim or add another spoke to fill in the spot, if necessary. Weave in these areas with materials such as Coconut coir as it is soft and will pack well.

TIP: I finished my basket by adding small pieces of philodendron sheaths to create an extra woven area between the antler points (see photograph). Experiment and see what you can do with your antlers.

Mini Antler Basket

Judy Briscoe

Materials

- Antler tip or very small spike
- #3 smoked or natural round reed
- Dyed reed, size #000 or #00
- Assorted yarns and waxed linen
- Fine tips from palm inflorescence

Drilling

- Drill one hole in the heaviest end of the antler and then drill a second hole about 1/4" in from the first hole.
- Drill two more holes in the smaller tip end of the antler. Drill at least a 1/2" in from the very tip.
- Finally, drill two holes on the underside of the antler, one even with the front tip hole and one even with the opposite end hole.

Frame and Weave

Using a length of #3 round reed, determine the shape of the basket by inserting the reed through one of the heavy ends, then through one of the corresponding point holes, and back to the heavy end. Trim and glue. Do the same for the second rim, making sure to keep the spacing even all the way around, about 1/4". You have now created two basket rims, one inside (smaller rim) and one outside. Using another piece of #3 round reed, determine the depth of the basket, and then trim and glue in the underneath holes. This is your center spoke.

Begin weaving on the heavy end of the antler with a length of either #000 or #00 fine round reed. Start with the end of the weaver underneath the center spoke. Bring the weaver up and over the first outside rim, then under the inside rim, and over and back underneath the outside rim. This will now cause you to weave over the center spoke to the opposite side of the center spoke, where you will go under the outside rim, over the inside, around and back over the outside rim, and under the center spoke. Repeat for about 5 to 6 rows. Don't cut the weaver. Go to the tip end of the antler and repeat this process with a new weaver.

Start adding spokes by trimming the ends of lengths of #3 reed and inserting them in the spaces created by the over/under weaving. Fan out these new spokes so that they start to create the sides and bottom of the basket.

Weave another 1/2 to 3/4" at each end of the antler, going over/under each new spoke to create a continuous weave from one side of the basket to the other. Weave both ends of the basket. Add more spokes, usually four at this point — two on either side and two at the bottom — on either side of the center spoke.

Change weavers and use waxed linen or yarn to hold these new spokes in place. Go back to the fine weaver for another 6 to 7 rows; add new spokes as needed. Continue with the fine reed weaving to hold the spokes in place, and then switch to a softer material like yarn for texture. You are still weaving one side and then the other.

Once all of the spokes are in place and the basket is shaped and strong, add the palm inflorescence to the very bottom, weaving from one side of the bottom of the basket to the other. Use a packing technique if needed. This means if you are running out of rim space, yet you still have a lot of spokes to weave, turn on the spoke below the rim instead of going to the rim to turn. Do the same on the opposite side of the basket. In the next row of weaving, turn on the second spoke from the rim. This allows you to "pack" the weaving without creating a "V" at the bottom of the basket. Let dry and add more palm if needed.

Author's Note: As a general rule, it's best not to use soft materials to hold spokes in place. However, with this style of very small basket, which will not have any stress applied to the finished product, it is acceptable to use waxed linen or other soft materials to hold the spokes.

Honeysuckle Antler Basket

Judy Briscoe

Materials

- Fallow deer antler
- #6 smoked reed
- Honeysuckle, stripped of bark
- Red and yellow cedar
- Sweetgrass
- Natural rush
- Coconut coir
- Palm inflorescence (Queen palm)
- Other natural fibers
- 3/16th bullet bit and a drill

Drilling

Drill one hole in the shed end of the antler, 2-1/2 to 3" from the end of the antler, and insert a length of #6 round reed into the hole and bring towards the point end to determine the shape of your basket. Mark where the holes should be drilled in the point end; again they should be about 2-1/2 to 3" in. Drill this hole. Now, drill two more holes in the shed end and two more in the point end, all about 1/2" from each other. Insert reed into each set of corresponding holes (end and tip) and create a shape with three rims. Glue the ends into the shed end.

Next, drill one hole at the underneath side of the shed end to create the bottom spoke. Glue this spoke in place; shape the bottom (not too tall or it will not sit) and tie off the other end of the spoke to the tip end of the antler. Use waxed linen or whatever is handy.

Weaving the Basket

Begin the weaving at the shed end of the antler. Start with the smallest strand of wet honeysuckle. Start behind the middle bottom spoke. Weave over and under until you reach the inside rim spoke, and then turn and weave back to the other side of the basket. Reverse direction, weaving back to the opposite side. This forms a fan shape so that you can add spokes. Weave about 3/4 to 1", leave the weaver hanging, and repeat with a fresh length at the point end of the basket. At the point end, weave only over the spoke and rims, not the antler.

Now it is time to add spokes. Add one on each side of the bottom spoke. These will typically be 1/2" longer than the middle spoke so that when they are woven in place they will spread out, creating the bottom of the basket. Add one spoke to each side of the basket and underneath the rim, again about 1/2" longer than the rim as these, too, will spread downward and create the basket sides. Pick up the strands of honeysuckle and weave these new spokes in place. Weave (add new weavers if needed) about another 1/2" on both ends of the basket.

Insert six more spokes in the spaces created by the weaving. One each on either side of the bottom spoke, one on either side of the two you added at the bottom, and one on each side underneath the spokes added under the rim. Continue weaving with the honeysuckle for at least another 1" so that these six new spokes are held in place.

You may now change color or texture. Add new spokes as needed. The rule is that there should be no more than two-fingers width between spokes at the very bottom of the basket. Here are some guidelines to go by when adding other textures:

- Sweetgrass should be unbraided and soaked for 30 to 60 minutes before weaving. Use two bundles of five or six strands with a twining technique rather than straight over/under. When a strand gets too short, add another piece.
- Red cedar is cut 3/16th wide and should soak until soft, at least 15 minutes. Weave it over/under.
- Yellow cedar is cut 1/8" and should soak at least 15 minutes or until soft. It is used as a twining material.
- Natural rush is woven dry. If it's too stiff, spray it lightly with water.
- Coconut fiber is woven dry. Use pieces over/under no longer than two yards or it will tangle.
- Palm inflorescence (used best at the bottom for strength) must be soaked overnight, or at least six hours, for it to become flexible. Start weaving with the heavy end to the thin end and repeat as needed. If used at the bottom of the basket, check the tightness of the weave the next day, as the inflorescence may shrink slightly in the drying process. More strands might need to be added.

Moose Antler Basket

Fonda Haddad

Materials

- Moose Antler
- Minwax® (I used Dark Walnut)
- #8 or #10 round reed
- 1/4" flat reed
- 3/8" half round reed
- #2 and #4 round reed
- #3 Rebar, rust removed, and enamel paint
- Glue (Titebond® II)
- Beads and cordage as decorations

Moose are the largest animals of the deer family and can weigh up to 1,000 pounds. Their antlers are very large and heavy, so extra planning must be done to accommodate their weight. The antler used in this project has seven points; the moose was killed in Alaska by my brother, Alan.

First, clean the antler with steel wool and then stain it with the Minwax®. I like the Dark Walnut stain.

Next, begin by envisioning the shape of the basket. The antler is the handle so it will be up in the air. You must decide on where the spoke will go so they will support the heavy antler. I decided for this basket to use an Appalachian Hen Basket design.

Drilling for Spokes

Begin by drilling holes with a 9/32nd drill bit in each of the points of the antler. The holes are angled to allow the spoke to travel from underneath the point to the butt or head end (the end that is attached to the head). At the butt end, because there is not enough room to drill seven holes, I drill two lines of holes about 1/2" away from each other: one with three holes and one with four holes.

Working with the heavy #10 reed, open the pound of coil and allow the reed to relax a little. I like to roll the reed between my hands, too, to help straighten it out. Misting it with water also helps. Cut the spoke length to fit the imagined space between the points and the head end, and add a few inches for good measure. Sharpen the ends of the spokes by putting them in a pencil sharpener, or use a knife and trim them. The outside spokes should be longer than the ones that create the basket bottom. Again, you need to imagine the shape of the basket with weaving on it and shape the spokes accordingly.

Glue the spokes into the point end of the antler. Let the glue dry for three to four hours. Only then do you bend and secure the remaining spoke ends to the head end of the antler, but be beware of the fact that they won't stay in place. They need to be glued and then tied in place using heavy cord or twine across the middle of the bent spokes. This helps control them and keeps them in place during the 24-hour drying process.

Weaving the Basket

Before doing the actual weaving, prepare additional spokes. Cut six new spokes longer than necessary. Trim one end. Tie this trimmed end to the glued spokes at the head or butt end only, about 2" down from where the first seven are glued. Keep the total number of spokes odd. Let the untied ends of these new spokes loose for now. They will be picked up later in the weaving on the point side of the basket.

Using damp #1 round reed, begin the over/under weave around the original seven glued spokes at the point end for about 2". Add the new spokes into the weaving, fanning them between the original seven; make sure the shape is kept outward, not pulled inward.

Now, for the point end of the basket: Using 1/4" flat reed, weave over/under each of the glued spokes and the ends of the six new spokes. Adjust their length so they help create the shape you want. Weave for a few inches until all the spokes are secure. The extra spoke material at the top of the six new spokes can be trimmed later or hidden behind a decorative addition of leather and beading.

Keep adding spokes as needed, but keep in mind they should be no more than about 2" apart from each other. Add weavers as needed and switch from round to flat as desired. I started with seven spokes and eventually ended up with a divided basket. The main body, which is flat on the bottom, used the five largest points. The mini-hanging basket used the other two points. All in all, I used 23 spokes. This number will vary each time a basket is woven, as it depends upon the size of the antler and the number of points.

Continue weaving as desired: I ended at the bottom with 3/8" flat oval because I knew I'd need to insert the weights into the basket to help it remain upright.

Bottom Weight

This basket was obviously antler-heavy, so I needed to weigh down the base to keep it from rolling over. I created a band of weights by using #3 rebar, cut into six 8-inch pieces. The rebar must be wiped free of rust, submerged into rust remover and sealer, and then sprayed with enamel paint to keep the rust from coming back. I also glued cotton cordage onto the rebar by laying down a coating of white glue and wrapping the cordage around the rebar. I then secured it with rubber bands until dry.

Once the rebar was ready, I inserted the lengths into the weaving at the base of the basket after first misting the reed. Then I weighted down the bottom and let the whole area dry overnight.

Finishing Touches

Once the basket was dry, I trimmed off any excess spoke material. Where I could, I added a large wooden bead to one of the freestanding spokes to keep the weaving in place. In other areas, I added a few "false" spokes and wove a pattern up into the curve. I also used leather cordage to tie off the spoke end, again to keep the weaving in place. Additional beads and fish weights were added, hanging down from the head anchor for decoration and weight. A metal moose was engraved with my name and attached among the beads.

Inserting rebar into the basket bottom helps weigh and stabilize the very large, top-heavy moose antler.

Random Weave Antler Baskets

Karen Zane

Materials

- Small deer antler
- #5 or #6 round reed
- #1 round reed
- Drill and bit to fit the #5 or #6 reed
- Glue gun
- Small awl and scissors

The random weave basket makes good use of small antlers. With just a simple rim and spine frame you can make an art piece in an hour. They make a nice cardholder or accent piece for your bookshelf.

With colored reed, you can create your own taste by using one color or several. Small beads woven in give a nice accent to the basket. Added embellishments bring out the charm in each basket.

This is a great basket to let your imagination run wild without involving too much time or cost.

Framing the Basket

Even though we will "random" weave, the basic basket structure still needs to apply. Start by choosing a small antler and drill holes in the butt end and the point end to accept a length of #5 or #6 reed. The size of reed used will be determined by the overall antler size. Thread through a length of reed to create the basket's rim and hot glue the ends into the butt end.

Next, drill two holes for the middle rib or spoke. The first hole is typically underneath the butt end with another shallower hole underneath the point end. Glue this spoke into position. You now have three ribs or spokes on which to start the weave.

Random weave

Using a damp length of #1 round reed, begin at the middle spoke at the butt end of the antler. Place the tail of the weaver against this rib and using an over/under weave, go from rim to rim, double wrapping at each rim. Catch the tail of the weaver so that it is secure. Do not put pressure on the rim — just hold it gently in place and weave with a light touch so as not to distort the shape of the basket. Weave for four to five rows at the butt end only. Stop weaving after a double turn on either side of the basket rim.

Now, bring the weaver across the bottom of the basket to the opposite side, but at the point end of the basket. This should have you about 2" away from the point where the reed enters the antler point, but diagonally across from the butt end rim. Wrap the weaver twice around the rim. Repeat, bringing the weaver back and across from rim to rim, always wrapping twice over the rim. Continue in this manner, using an awl to help feed the weaver through (over and under) some of the previous crossings. To help tighten up the weaving, make circles that will pick up loose weaving. To add new weavers, lay the end of the new weaver alongside the tail of the old weaver and weave in and out beside each other for a few stitches. Eventually, the ends can be trimmed or worked into the weaving.

If the weavers start to get bunched up, separate them by weaving through the bundle. As the weaving gets tighter, you will be able to move the woven places and get the effect that you want. There are no mistakes in this type of basket weaving, so have fun. Keep filling the basket until you get the shape and effect you want. The frame should be in the shape it started and the weaving tight, allowing the basket to sit upright or however you intended. You can stain the basket if desired and add embellishments.

Medicine Woman

Medicine Woman is a collaboration between my stepson, J. Scott Begley, and I. The fallow deer antler just begged for some artwork, and Jody loves Indian art. What better combination could I ask for? With inspiration from Jill Choate's work, I connected the one fallow antler with three whitetails to make a circle. After drilling the antlers and weaving the basket, Jody began his story.

It's a tale of an Indian woman whose Medicine Man husband had died and what happened when the tribe subsequently shunned her. The hunting had dried up, and the tribe was in danger. The woman used the powers she had learned from her husband to call the animals. As a result, the tribe accepted her as their Medicine Woman — she became the first such woman in their tribe.

The age cracks drawn on the antlers gives the impression of an electrical charge given off by the power of the Medicine Woman. The circle (created by the four antlers) envelops the tribe and protects their land. The colors of brown earth and teal sky bring together the forces that create our world. The twining of Japanese cane and waxed linen represent the unity needed for our survival.

I am very proud to share with you our collaboration.

Close-up of the fallow antler used with this basket tells part of the story, created by J. Scott Begley, on life's trials and triumphs.

Chapter Six

Gourds

Antler Gourd Basket

Nan Noble

Materials

- Round gourd of any size
- Deer antler big enough to span the gourd
- Thin deer hide, preferably with a ratty edge
- Acrylic paint (I used three shades of brown plus metallic)
- White craft glue and flat black spray paint
- Painting materials (sponges, pallet)

Preparing the Gourd

Most of you may know how to prepare a gourd for use, but let's review:

- Soak the dirty gourd in water or a wet towel for about an hour.
- Using a household copper scrubbie and cleanser, gently remove all the field dirt and mildew/mold. Both Comet® and Ajax® cleansers contain bleach that helps combat the mold and mildew.
- Let the gourd dry.
- Pencil mark the opening cut (wavy is more effective than a plain cut for this gourd).
- Using a hand or mini-jig saw, cut away the gourd top.
- Using a cleaning tool (ball cleaner on a power drill, grapefruit spoon, gourd scraper, etc.), remove the interior membrane and seeds, and then sand the inside until smooth.
- I like to use flat black spray paint to finish off the inside of the gourd.

Painting the Gourd

- On a pallet or paper plate, squeeze out a small dab of acrylic paint in each shade selected, and then dip a small piece of sponge into water and squeeze it as dry as possible.
- Dip the damp sponge into the darkest paint and blot onto a clean area of the plate to remove the excess.
- Pat the sponge over the entire gourd, letting some of the natural color show through.
- Reapply paint to the sponge as necessary; be sure to always blot the excess off.
- Repeat this process, using the middle shade and then the lightest.
- Finally add some metallic using the same process; gold, copper, silver, or bronze will do.

Applying the Leather

Choose a piece of leather that is long enough to go around the top of the gourd and will also overlap the rim both inside and out. I prefer leather that has a lot of holes and scrappy edges, but smooth leather may also be used. Start by putting a bead of glue close to the top edge of the gourd exterior, about 1/3rd of the way around at a time. Carefully glue on the leather to the outside of the gourd.

Once the exterior is glued in place, apply glue to the underneath side of the leather that will flop over the rim. Be sure to do this just a little dab at a time. Pinch and push the leather into an interesting shape as you glue. Trim off any extra from the inside of the gourd to about 1/2". Glue any portions of the inside flap that remain loose.

Attaching the Antler

Lay the antler over the gourd until you find the best position. The antler should have at least two or three points that touch the gourd rim. Now you have to determine if you can drill through the antler and the gourd to lash the two together, or only drill the gourd because the antler tips are too small. Be careful drilling the antler tips, as they tend to be dry and can easily crack. The holes should be about 1/4" in diameter, large enough to accept the leather lashing. This can be purchased leather lashing or you can cut one about 12" long from the left over rim treatment.

If you can drill the antler, bring one end of the lashing out through the gourd hole and the other end through the hole in the antler. Tie the two together in a double knot. Or if desired, drill two holes in the gourd and bring both lashing ends out to the front, then one end through the hole in the antler, and tie off. If you can't drill the antler, simply bring the lashing out from the gourd through one or two holes and lash over the antler. If needed, use a small dab of glue where the antler hits the gourd rim and cover this with the lashing.

Coiled Gourd with Antler

Laraine Short ~ Deer design by Cheryl Dow

Materials

- Pine needles, dark brown
- Tagua nut
- DecoArt® Americana Matte Spray and Bittersweet Chocolate acrylic paint
- Tandy EcoFlo Bison Brown
- Tapestry needle, size 20
- Antler, 11" long x 8" wide
- Wash brush and water basin
- Waxed linen, dark brown
- Woodburner, small ball tip
- Hair dryer
- Drill and 1/16th and 3/64th bits
- Scissors and Ruler
- Scraps of white knit fabric
- Assorted bone beads and embellishments

Preparing the Gourd

For this project, we used a large gourd with an angled opening, one that would allow the antler to rest within the coiling and create a decorative look to the opening. Ultimately, we will bring all the coiling up to a small opening at the top of the gourd, where we will end the coiling at the tagua nut.

Using the 1/16th drill bit, drill holes around the edge of the cleaned, opened gourd — 1/4" from the edge and 1/2" apart. Paint the inside of the gourd with wash brush and bittersweet chocolate paint. Allow it to dry or use a hair dryer. In a circular motion, using a piece of fabric scrap, rub on the Bison Brown dye and let it dry. Finally, spray the gourd with matte acrylic spray.

Coiling and Adding the Antler

Start sewing dampened pine needles to the gourd, two per hole, with their caps still on. Work to the left. The stitch comes over the gourd rim. Once you have gone all the way around, pick up the previous row's stitch to secure each new row of coiling, adding pine needles as needed. Remove the caps from the needles once you have done the first row. Determine where the antler will best fit and work one row of coiling over the antler and then under in the next, securing the antler into the coiling.

As you can see in the accompanying photo, the coiling does some free-form after the first few rows, working up the gourd and over/around the antler, until the coiling gets smaller and smaller. The opening at the end should just fit the tagua nut.

Once all of the coiling is done, use a hair dryer to dry the needles and wipe down with a fabric scrap. This will remove any loose wax from the waxed linen and add sheen to the pine needles. Thread beads and add to the coiling where there are any wide spaces.

Boil the tagua nut before woodburning the design. This makes the nut harder and therefore easier to woodburn. Using the 3/64th bit, drill two holes through the tagua nut. These will allow waxed linen to be used to tie the nut to the coiling. Woodburn a design on the nut and then attach it in the small opening in the coiling.

You might need to add rocks to the inside of the gourd to offset the balance of the antler.

The coiling, which goes over and around the antler, becomes smaller and smaller at the opening in the gourd until it ends at the woodburned focal point, the tagua nut.

Leather Rimmed Gourd

Mary Gayle VanIngen

Materials

- Cleaned gourd
- Antler that suits the size and curvature of the gourd
- Black acrylic paint
- 1" paintbrush and rag
- Wood stain
- Spray acrylic sealer
- 1/8" suede lacing equal to six times the circumference of the rim
- 4-ply waxed linen
- Tapestry needle and awl
- Dremel® and bit, or other power tool

Staining and Weaving

Prepare the gourd by first painting or spraying the interior with black acrylic paint. Let it dry well before handling. Then using whatever wood stain suits you, apply it to the outside of the gourd with a soft cloth or brush; wipe down and let dry. Repeat this process if a deeper color is desired. Finally, spray the outside of the gourd with spray acrylic sealer. It's best to use three to four light coats so as to not cause drips.

To start work on the leather lacing, first drill holes around the rim of the gourd. You can also use an awl, if preferred. The holes need to be large enough to fit the tapestry needle and the waxed linen. Make them 5/8" apart and 3/4" down from the rim.

Thread a long length of waxed linen into the tapestry needle and proceed to stitch over the rim all the way around the gourd. Start by inserting the needle through any hole, from the outside, leaving about a 4-inch end. Using an overcast stitch, go all the way around to the beginning. Keep the stitches tight. Tie off the two threads, but do not cut the thread that is attached to the needle.

Taper one end of the leather or suede lacing and lay it horizontally along the rim, with the waxed linen (still on the needle) laying between the rim and the leather. Hide the knot as best you can. Now, go over, around, and under the leather, and then under the waxed linen overcast the stitch that is ahead of the needle. Continue this process until you return to the beginning.

Start the second row by positioning the thread between the two rows of leather and continue as before. Stitch on as many rows as you can, but be sure to end up back at the start point, ending the leather by tapering it and over-stitching it to the previous row. Tie off the thread by weaving the needle through the weaving until you can pick up the original 4-inch end. This leather lacing creates a soft look to the rim that contrasts nicely with the hard, curved look of the antler.

To secure the antler, assess the shape of the antler and the gourd and mark where the butt end or points lay on the gourd rim. Carefully drill holes (small enough to allow waxed linen to be threaded through) in the points and butt end. Be sure to take care when drilling at the points, as they tend to be drier than the rest of the antler. Thread the waxed linen through the holes and then into the weaving, and tie off. Trim the ends.

After the leather rim was stitched in place, the antler was carefully drilled at the point and butt ends and lashed to the rim.

Chapter Seven

Wall Hangings

Gourd Masks with Antlers

Bonnie Gibson with photography and copyright by Everett Gibson

Creating gourd masks is always an adventure. What do I have in my studio that could be used? Feathers? Cactus ribs? Turquoise cabochons? All of these can prove helpful, along with my favorite acrylics and oils, to make one-of-a-kind masks.

For me, the most important thing is to make sure the antlers are securely fastened to the gourd. I often drill holes in the end of an antler and glue in a bolt with the head cut off. You can cut the ends off with a hacksaw or one of the thin cutting discs that are in most Dremel® kits. I like epoxy for a strong bond. Then I drill a hole in the gourd and use nuts to tighten the whole thing down. You can also use washers on the inside of the gourd to distribute the antler weight.

Textured paint and a raffia fringe were used on one of the masks shown here. The tongue was carved pine and the frame was created from saguaro cactus ribs. Another has nosepieces made out of gourd necks. Several have inlaid turquoise cabochons.

When I need antler slices, I cut my own on a bandsaw. The slices resemble buttons and can be sewn on or hung from leather fringe. If you are going to cut your own slices, be sure to wear proper respiratory protection, as deer antler dust is very fine and the smell is strong.

To add feathers, I used a base of Makin's Clay® and/or holes drilled directly into the gourd, especially for very large feathers. Again, I like to make sure even the feathers are securely fastened to the gourd mask.

Deer Horn Mask

This "Saguaro Frame Mask" utilizes cactus ribs for the framework.

Single Prong Antler Mask.

Antler Wall Basket

Bob Hosea
Photography by Brittany Hosea

Materials

- Deer antler
- Seagrass, baby or #1
- Round reed for spokes, #4 or #5
- Dyed flat 3/16th reed
- Drill and bit to fit spoke size
- Glue and shears

Wall antler baskets can be created from any deer antler, but the best is one that will allow the basket to rest flat against the wall. Typically, the outside curved part of the wall basket antler is towards the viewer with the points towards the wall. Select the best one for this project and proceed. The spoke material size will also depend on the overall size of the antler: The larger the antler, the larger and stronger the spoke material. This style of wall basket is woven in two steps: first the back as it lays against the wall and then additional drilling and spokes are added to create the front.

Drilling the Antler

Using the appropriate bit for your spoke material, drill two holes on opposite sides of the base of the antler (butt end). The hole need not go all the way through. Insert into one hole a length of spoke material and bring the reed around to the opposite side, creating the shape of the rim. Once you are satisfied with the shape (the rim should also touch each of the antler spokes on the way by), trim off any excess material and glue in both ends of the rim material into the holes.

Next, drill two small holes in the antler tips and tie these tips to the rim. Take care when drilling the antler tips for they are usually dry and can split easily. Don't twist the drill or bit as you create this small hole and go at least 1/2" up from the very end.

Once the glue has dried and you have secured the rim to the antler tips, drill three more holes for additional spokes. These three holes are on the backside of the antler, where it will face the wall. They are also in line between the two original holes.

Cut three more spokes, and then glue them into the butt end. Use round nosed pliers to pinch the ribs at the butt end and carefully bend the spokes up towards the tip end of the antler. Tie together these three new spokes at a point just above the original rim.

The framed antler is ready for weaving. The main rim is glued into position with the three additional spokes glued at one end only. The opposite ends are tied together.

Weaving the Back

You are now ready to weave both the top and bottom of the wall basket's backside. I like to weave a few inches at one end, then the other, and meet in the middle. This helps create a more even shape and a uniform weave. Start at either the top or bottom with sea grass in a simple over/under continuous weave. I like to start with this material because it is flexible and easy to use. You may use whatever small weaving material you like. Weave for about 2" at either end and add several more spokes. The total number of spokes added will depend on the size of your antler, of course. I needed a total of seven, including the original rim.

Initial weaving with two additional spokes in place. The number of spokes added will depend on the size of the antler.

Once all of my spokes are in place and woven into the back, I switch to flat dyed reed for a texture and color change. Then I end up in the middle with more seagrass. You are now ready to add more spokes and weave the front of the basket.

The entire back of this antler wall basket has been woven.
The basket will now be flipped over and the front side created.

Weaving the Basket Front

For this basket, I determined that three spokes on each side would be enough to create the basket front. You may need more depending on the antler used. I drilled three holes along the sides of the antler and three more along the point sides to accept the new spokes.

I chose to use more of the dyed 3/16th flat reed to continue with the color pattern on the back. These two sides of the front are independent so they can be woven separately, but as you are weaving back and forth, when you reach the original rim, thread each weaver between the already-completed rows of weaving for the back. Wrap the new weaver around the rim to create an interlocking of the front to the back. Continue until you have woven each side, changing materials for texture and color if desired.

Once all of the weaving is completed, trim any loose ends of the weavers and trim the long ends of the original three primary ribs on the backside of the basket. If desired, a small amulet can be attached to the basket at the center of the fork in the antler.

Three more spokes have been added on each side of the antler, creating the front of the basket.

Bob has woven through the original basket rim as well as the three new spokes on each side of the antler. He chose to weave with the same material that was used on the back of the basket.

Antler Wall Pouch

Betsey Sloan

Materials

- Curved antler, preferably whitetail
- #6 and #5 smoked round reed
- Waxed linen
- #2 smoked round reed
- #3 dyed round reed
- Palm inflorescence
- Coconut coir, seagrass, or other natural materials
- Drill and bit to fit #6 smoked reed

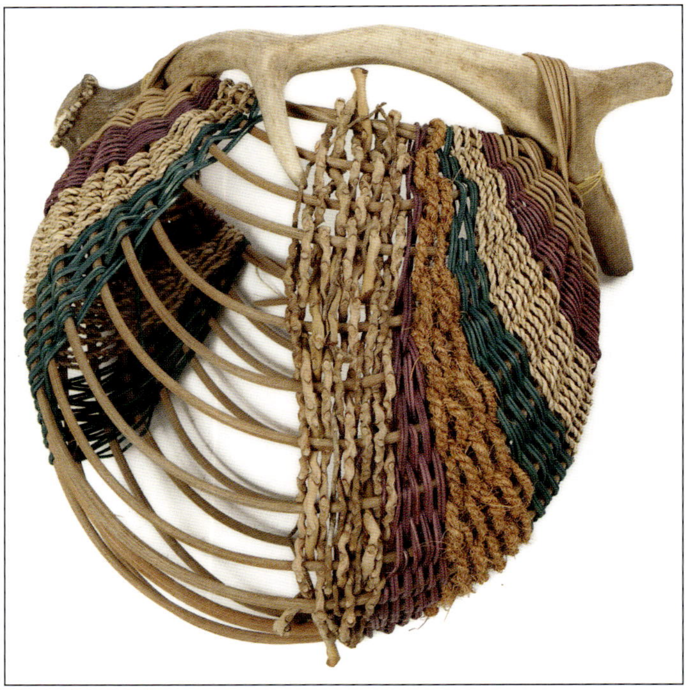

The wall pouch has not been finished in order to show the number of spokes required to shape the basket on the antler. The remaining space will be filled on the left with matching materials and the center filled with soaked palm inflorescence.

Drilling and Framing

Drill two holes in the bottom of the antler for the center spoke. In a curved whitetail, one hole is at the butt end and the other at the point located at the innermost curve of the antler. See the partially completed basket with this project for positioning, or read more about drilling for antler wall baskets in the chapter on drilling.

Feed a length of #6 reed from hole to hole and create the basket's center, or lowest point hanging down from the antler. The length of the center spoke will depend on the overall design desired. A too long or too short spoke will just not look right. Glue the spoke in place.

Create a "D" ring from soaked #5 reed. Take a length of reed and twist it around itself two times, creating a ring. Bend the ring to create a "D" and tie this ring to the antler. The flat portion of the "D" is the back of the basket. The curved portion fits snugly against the curve of the antler. Waxed linen is used to tie the corners to the center spoke and the antler.

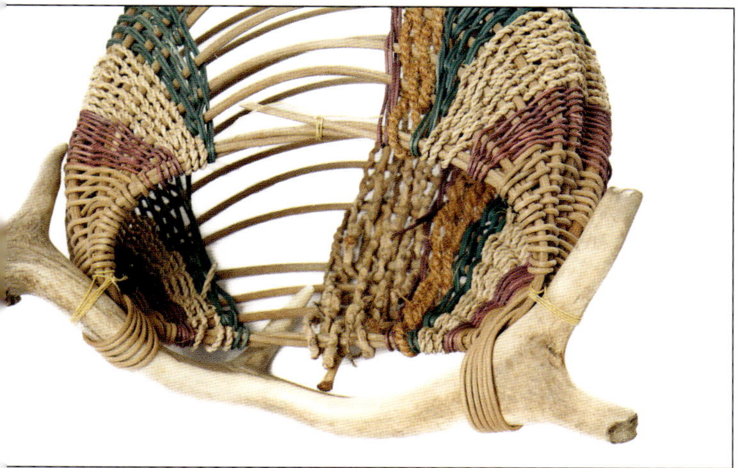

A "D" ring was shaped and tied to the center spokes and the antler. This ring rarely keeps its shape and ends up more rounded, like a natural antler basket rim.

Weaving

You now have three spokes to weave on at each side; the back rim, the center spoke, and the front rim/antler. The beginning weave for this basket is woven around the antler and the front rim as one unit. Starting with soaked #2 smoked reed at the underside of the center spoke on the left side of the basket (the basket will be facing you), start an over/under weave. Go over the antler/spoke together at the front of the basket, bring the reed around through the top of the basket and down to the inside, and then weave over the center spoke and on to the back rim. Go under the back rim and reverse direction to an over-weave. Weave under the center spoke and then over the antler/rim at the front; reverse direction. Repeat for about 1". Drop the weaver (do not cut it) and go to the right side of the basket.

Soak another length of #2 smoked, but start OVER the center spoke, then under the back rim, back to the center for an under, and to the front rim/antler for an over. In this way, the weaving is the same at both sides of the antler, with an over-weave of the antler/rim at the front. Continue the over/under weave for about 1".

Cut some lengths of #6 round reed and insert new spokes into the spaces created by the weaving. Their positioning should be such that the basket or pouch begins to fill in the spaces between the center spoke and the antler/rim. Two of the spokes should be almost the same length as the bottom spoke. I like to do this so that the pouch actually has a slight flat bottom to it. This is beneficial when adding the final palm inflorescence to the center. See the photo with this project for correct spoke placement.

Pick up the ends of the #2 weavers and weave the new spokes in place. You may now go over/under just the rim at the front of the basket. Stop weaving over the antler as part of the rim. When the weavers are finished, add more spokes if possible. Start with another weaver, either soft or hard. However, be sure the new spoke ends are inserted into a hard material or they will eventually work their way out of the weaving.

Note the over/under weaving for the first few rows goes over both the antler and the rim, as one unit, at the wall basket's front side Only after the first few spokes are added does the weaving then go over the rim only.

You will need to start packing almost immediately once you start the second weaver. To better understand packing, read that chapter. Basically, you will need to turn on spokes other than the rim in order to fill in the sides of the pouch so you are left with a desired straight swath at the very center of the basket, which will be filled in last. Never turn on the same spoke twice in a row, but you can turn on the spoke again if there is one row of weaving between the turns.

Continue adding spokes and change materials as needed, weaving one side of the basket and then the other with the same material. Once you have woven so that the pouch is a center section, switch to very well soaked palm inflorescence or other filler material and weave from one side of the pouch to the other. Pack where needed.

A soaked philodendron sheath may be woven in for texture, as well as beads and feathers. I like to finish the basket with a dried pod tucked into the points of the antler or at the very center of the front as a focal point. Add a length of leather lacing, tied to the back spoke, for hanging the pouch.

Antler and Gourd Wall Basket

Marianne Barnes

Materials

- Half of a canteen gourd
- Small antler
- Leather, beads, and wood for backing
- Acrylic paint and clear acrylic sealer
- Dichoric glass
- Woodburner and tips
- Oil pencils

Using a medium-sized canteen gourd, cut it vertically through the center. Clean one of the halves using a ball cleaner or scraper, and then sand it smooth.

Cut a piece of wood to fit the shape of the half-gourd. This will create the back of the piece. If wood is not available, use any material that will remain rigid.

Determine the top of the basket and lay the antler along the top most edge of the gourd. The antler curve should fit snugly to the gourd. See accompanying photo for placement. Draw a pencil line around the curve and cut out the gourd to match the antler's curve. This gives a scooped out look to the front of the gourd.

Glue the back material to the gourd using wood glue. If using wood, also use small nails or tacks at the top and bottom. Paint the inside of the gourd and the backing. I used dark brown acrylic.

Drill holes in the front of the gourd large enough for leather lashing to hold the antler in place. Carve out the area in the center front (most likely where the stem was attached to the gourd) to hold the dichroic glass. Also drill two holes in the back material, at the top, for a hanger, which will be leather lacing.

Draw on the dragonfly designs and woodburn them, and then woodburn lines around the front of the top opening and around the carved area for the glass as decorative borders. Use oil pencils to color the dragonflies and then acrylic sealer over the entire gourd.

Glue the dichroic glass in place. Using leather lacing, attach the antler to the gourd front rim and create a hanger with more lacing for the back of the project. Add beads, feathers and whatnot as desired.

Chapter Eight

Other Creative Projects

Antler Bowl

Betsey Sloan

Materials

- 8 to 9" curved whitetail antler
- #4 smoked round reed
- #2 and #3 round dyed reed
- Baby Seagrass
- Pods as embellishments
- Drill and bit to fit #4 reed

Drilling

In the chapter on drilling, where and at what angle to drill for an antler bowl is covered. In short, here is what needs to be accomplished:

- Determine first where the bowl/basket will be woven up from the antler. Find the best way that the antler sits on the table.
- Using a pencil and starting at the butt end, draw an elongated circle along one side of the antler (the edge of the curve that is facing upwards), then around the curve by the points (but not on the points), and then back along the antler where it touches the table, the butt end. This is the drill line.

- Mark dots about 1/2" apart and, using a drill bit that fit's the #4 round reed, drill holes for the spokes. These need to be deep enough to hold the reed; don't drill all the way through the antler. Be sure to angle the holes so the spokes can flow up along the points, lay flat at the bottom to create the basket base, and pouch out at the butt end. Some of this will come with practice — the being able to "picture" the bowl/basket on the antler and then drilling for the spokes on that angle.
- Using a good waterproof glue (I like Super Glue Gel), glue in the spokes.
- Cut the spokes for as tall as you want the basket PLUS 10 inches.

Weaving

I like to start with a soft material, such as funky yarn or baby seagrass, as it helps cover the drill/glue holes. Twine for about 6 rows or more if you are using yarn. Switch to any dyed or natural round reed, preferably #2 and twine. The number of rows depends on your own personal choice. Be sure to help the spokes flow up the antler points, and weave so the butt end spokes are able to pouch outwards, creating that side of the bowl. A tight weaver may have trouble with this style of antler basket. You need to have your weaving tight enough to look good, but not so tight as to pull all the spokes inward, creating more of an upright cup rather than a bowl. Switch materials as needed. Create a basket bottom by selected spoke weaving, though the final weaving must be a reed. Do not use a soft material such as yarn at the very top, under the rim treatment. It will tangle on you as you execute the rim.

Top Rim Weaving

I used a Japanese Rolled Border as a rim treatment. Soak the spoke tops in water until they are very pliable. Just turn the entire antler bowl upside down, submerge the spokes up to the last row of weaving, and weight down for 10 to 20 minutes.

Starting anywhere on the rim, take one spoke and pass it to the right, behind two spokes, and then out to the front. Take the next upright spoke to the right of that first turned spoke, and then take it to the right and behind two spokes and out. Repeat all the way around the rim. When you get back to the starting point, you will run out of places to "go behind." Simply push the folded over spokes back out and insert the last two in the correct spaces. All of the spokes should now be to the outside of the basket.

Starting anywhere, the second step for the rim is to pick three spokes. One is in the left hand, two in the right. Tug slightly on the farthest right spoke. You have created

a small opening next to that spoke. The spoke in your left hand goes over the two spokes in the right hand and into this opening in the weaving. Tug this spoke into the center of the basket. Move to the next spoke to the right; pick it up with your right hand and tug. With the left hand, take hold of the left spoke; go over two and into the space. Repeat all the way around the rim. When you get back to the start, figure out where the final two spokes should go by backing out the two folders over the spokes. All spokes should now be to the inside of the basket. Mist the spokes again if they are getting dry. The spokes must remain pliable all through this process. It's a good thing to not be distracted during this rim. It's easy to lose your place, never to be found again.

Here is the final step for the rim. Mark a spoke on the outside of the bowl as your start point. With your left hand, select three spokes and pull them free from any entanglements and out towards the center of the bowl. Using the right hand, take the bottom-most spoke in the series (the one by the "V" between the thumb and first finger) over the two spokes still being held, and then fold it down towards the bowl's bottom. Your left hand should remain inside the bowl, holding these folded spokes out of the way (with the little finger) while the remainder of the fingers keeps tension on the two remaining spokes.

Next, select the next spoke to the right; pull it out of any entanglement. You now have three spokes in your left hand again. With the right, take the bottommost spoke over the other two spokes and bend it down, catching it with the little finger. Repeat all the way around. When you get back to your start point (by the marker), determine where the remaining two spokes need to go by pushing out the first two folded-over spokes; insert the remaining two spokes and tighten. Once all of the spokes have been bent and inserted for one row, you may repeat this entire final step again. What happens is that the weaving "rolls over" into the inside of the bowl, finally hiding any trimmed ends. However, you may end the process after one "over and down" row, too. Trim the ends of the spokes (not too short as the reed may shrink slightly as it dries) and tuck the ends into the rolled border. You may feel like a contortionist while you are weaving this rim, but it makes a great finishing touch to antler bowls and baskets woven on a gourd. Just remember to mark your start point when you are ready to do the third step (over two and down) and talk to yourself it if helps. It does me.

I like to drill my embellishments in order to secure them to any of my baskets or gourds. Most of the pods and botanicals I sell online are hard enough to drill. Simply drill a small hole or holes in the pods and sew on with waxed linen. You are now finished.

Medicine Bowl

Betsey Sloan

Materials

- Fallow deer or similar palmated antler
- Cleaned gourd bowl with specially cut rim
- Tria® alcohol ink, brown
- Black leather dye
- Black spray paint
- Focal element, such as beaded feather embellishment
- Deft® spray finish
- Woodburner and tips
- Drill and bits
- Dowel, 1/4" diameter
- Leather lacing, beads, and feathers

I love working with fallow deer antlers. The large flat section at their point end is great for holding gourds or for use in woven antler baskets. For this project, I worked through several gourds and antlers until I found a pair that worked together. The antler needed to lie flat on the table and the gourd needed to sit on the palmated end of the antler, with at least two areas touching.

The first step for me is to drill the holes in both the gourd and antler to insure a good fit. I'd hate to create a great design and then blow it by messing up the drilling so that the gourd and antler don't mesh. I mark on the inside of the gourd bottom where it touches the antler in at least two areas. I then remove the gourd from the antler and drill through the gourd at one of these spots and replace the gourd on the antler, marking on the antler in pencil. Removing the gourd, I then drill the antler. The depth is about a 1/4" and the drill bit size matches the diameter of the dowel. Angle the hole in the antler as necessary. Fit about a 2" length of dowel into the antler hole. Place the gourd over the dowel. Make sure the connection is solid — gourd against antler. Now check to make sure the second pencil mark at the gourd inside bottom is where you want it to be. Sometimes after drilling the first connection, the gourd or the antler may shift position. Once you are satisfied, repeat the process. Drill the gourd, mark the antler, drill the antler, and insert another length of dowel. The end results should be a firm fit between the antler and the gourd. I glue the parts together after I've done my design and coloring.

If you mess up the drilling, you have two choices: ditch the gourd and start over, or re-drill the gourd or the antler and hide the mistake. QuikWood® or SpeedyBond® wood putty dries in about twenty minutes and can fill in an errant hole (so will any air-dry clay). The offending areas can be painted, stained, or hidden beneath leather lining at the gourd bottom. The antler hole can be filled also and, with a little work with wood stain or acrylics, carefully painted to match the rest of the antler.

My design is simply lines and woodburning. I made sure the rim of the gourd was cut to accommodate the round focal point of the bowl, the beaded feather embellishment purchased from Arizona Gourds. Using a compass, I drew the round area and then cut the gourd top accordingly. I figured out where the focal element would be glued to the gourd and woodburned a line design between that and the edges of the circle. Additional lines were drawn and then woodburned around the gourd, creating two areas, top and bottom.

I then spray-painted the inside of the gourd with black acrylic paint. I like to color the insides of my gourds and often use whatever is handy, from acrylics to leather dyes to Design Master® sprays. Let the inside dry completely before continuing. If you get some of the spray on the outside of the gourd rim, remove immediately with a paper towel and some mineral spirits.

Black leather dye was used on the bottom portion of the gourd bowl. The top portion was lightly stained with brown Tria®, mixed with a little alcohol to lighten the tone. I left the inside of the circle around the focal element unstained. Finally, I sprayed the gourd with Deft® satin spray acrylic, making sure I sprayed lightly for about five coats.

TIP: For those who use spray finishes, Deft® is available in hardware stores and is a terrific finish. It costs a little more than other spray finishes, but I swear there is more in the can. It is easy to spray and coats evenly. For those on a tight budget, most of your local superstores will carry a clear spray acrylic for under $1. Look in their spray-paint section if it's not in the craft section. This inexpensive spray works well too, but with any spray finish, the trick is to keep the can about three feet from the gourd and only use the "puff" end of the spray for coating. This is why about five coats are needed. Never allow the spray finish to be so heavy that it will drip. This results in a ruined gourd and I don't know of any way to repair dripping acrylic finish.

Okay, back to the project: Once the finish has dried, drill small holes on either side of the center focal point and string turquoise, red, white, and black beads on waxed linen. Using craft glue, affix the beaded feather embellishment to the front of the gourd. Allow the glue to dry.

Insert the dowels in the antler and place the gourd in position. Using a pencil, mark on the dowels where they should be trimmed at the inside bottom of the gourd. Remove the gourd, trim the dowels, and then glue them into the antler. Now glue the dowels into the gourd holes. Allow glue to dry and then stain or paint the dowel ends.

Because I used a fallow antler that had been trimmed off the deer, I wanted to take away from the cut mark. I added leather lacing, beads, and feathers as the final embellishment. To the inside of the gourd bowl, I added polished stones simply for interest. A sage bundle, small antler tip, and other items can be used, too.

Creating Floorvases

JoAnn Vanderheite

As a lifelong artist and gardener, I enjoy growing and using my own gourds for my art. Due to a short growing season where I live, I use these gourds for smaller pieces, therefore purchasing extra large gourds from southern growers. These large gourds provide wonderful forms with which I create my "floorvases."

In selecting a good gourd for these projects, I look for relatively flat bottoms for stability, good thickness for strength, and above all, those that have a pleasing and graceful shape. Since no two gourds are quite alike, their shear variety offers many opportunities for a unique design. Unusual necks, various heights, circumference, and occasionally severe but interesting blemishes can be the inspiration for an art piece. Thus the gourd itself can direct my artistic approach. At other times, I have sketched an idea and then searched for an appropriate gourd to suit the plan. With either method, the design must be fluid enough to change throughout the process in order to produce the most pleasing results. I have found myself led in new directions while working on a gourd and have become more flexible in allowing for that factor.

Although I work in various media, when I create art from gourds, their natural character seems to demand embellishment with equally natural materials. Antlers, twigs, seedpods, stones, feathers, seashells, and leather are some of my favorites. Antlers have a beautiful curvilinear affinity with the rounded shape of the gourds and I often use them for handles, trim, or sturdy bases. Antlers can be cut, painted, dyed, or left in their natural state. Due to their weight, antlers need to be anchored securely to the gourd, especially when used as handles. For this reason, I drill stabilizing holes or grooves into the gourd where the antler, or its prongs, will touch the surface. At these points, I use epoxy putty to firmly hold the antler in place. Even when completed, a gourd art piece retains its natural character and will sometimes respond to extremely cold temperatures or humidity, so a good bond is important. My intention is to produce a beautiful design that is strong and sturdy as well as aesthetically pleasing.

I have found that an extra large floor vase made from a gourd is not only a conversation piece, but works well in any interior where an accent is needed to decorate a dull area, such as next to a large piece of furniture or beside a fireplace or doorway. When finished, the color and design of the piece adds interest to a room and, when filled with natural branches or other dried botanicals, can lend additional height and visual impact.

Floorvase with Antler Rim

Using an extremely large gourd, this piece was designed to feature the three natural deer antlers that decorate the opening at the top. Using sweeping curves to draw the eye down the surface and reflect the curves of the antlers, I embellished areas with acrylic textured paint, slices of leather, and antiqued copper.

To anchor the heavy antlers, I first cut shallow areas on the rim where the antlers would rest naturally. Then I drilled holes in the body of the gourd in which to place the weight-bearing prongs or tips. I used epoxy putty to fuse the antler in place. The piece was then painted in areas of burnt umber and turquoise coloration in keeping with the natural feel of the design. The antlers were coated with a light coat of polyurethane. Natural turquoise discs were tied on with leather lacing. The piece measures approximately 18" tall by 18" wide.

Floorvase with three-antler rim.

Three whitetail antlers were secured to the top of this gourd by making slight grooves in the rim and then lashing the antlers in place after first using a slight amount of epoxy glue.

Gourd Vase with Antler Handle

The design of the piece was based on the form of the deer antler that was to be the handle. Using sweeping and abrupt curves for the design, I hoped to replicate the attitude of the antler in order to fuse the feel of the painted elements with the applied handle.

Using the button or butt end of the antler for the top of the handle, I hooked the antler over the rim of the gourd and cut a groove on the edge in which it would rest. For stability, three shallow holes were then drilled in the body of the gourd in which I set the first three prongs of the antler, leaving the last to float free. I anchored the antler in these resting places with epoxy putty.

Acrylic paint was used over both the smooth and textured areas of the design. The overall design and coloration were my stylized version of Southwestern pottery designs, with concho and leather fringe added for effect. The piece is approximately 20" tall and 12" wide.

Floorvase with antler handle.

This close-up shows the attachment of a very curved and painted whitetail antler. Holes have been drilled in the gourd's body to hold the antler tips in place.

Antler Pouch

Judy Briscoe

Materials

- Whitetail antler, curved
- #6 and #5 round reed
- #2 dyed round reed
- #3 round smoked reed
- Dyed sisal
- Natural vine
- Palm inflorescence
- Coconut coir
- Sugar palm
- Leather strips

Drilling

Locate the front curve in the tip end of the antler. Drill three holes underneath the antler, but not all the way through. These holes will shape the bottom of the basket. The front curve ends will be glued; the other ends will be tied off to the rim created in the next step.

Weaving

Create a ring of #5 round reed that will fit beneath the curve of the antler and also creates the opening on the pouch top. The ring is created by using a wet length of #5 reed and then twisting it around until the desired shape is achieved. The back curve of the ring should extend beyond the butt end of the antler, but should still be in proportion. Tie this ring to the antler.

Next, glue in the three bottom spokes of #6 round reed. Bring the unglued ends to the butt end of the antler and weave them using #2 dyed reed and a potato lashing to the basket ring. Continue using the same weaver, once the lashing is done, and weave over/under for about 1 inch. You might need to wrap twice over the rim every now and then in order to create the desired half-moon shape.

At the curved end of the basket, weave over/under the three bottom spokes and the basket rim, created by the ring. Do not weave on the antler itself.

Once you have sufficient weaving on each end of the basket, about 1", insert two spokes at the bottom (one on either side of the left/right bottom spoke) and one on each side of the basket, underneath the ring/rim. These should be about 1" longer than the existing spokes as they will spread out when woven and create the shape of the basket. Weave with the vine or #3 smoked reed for about another inch.

Add as many spokes now as possible, filling in the open space on each side of the basket. Securely insert these spokes into the "V" areas created by the previous weaving, and check the shape of the basket as you go. Looking at the photo, you can see that the sides actually drop almost straight down from the rim/ring. They don't flair out as drastically as spokes do with a standard antler basket. And the bottom spokes should lay flat on the table, creating the platform to securely hold the pouch upright.

Once all of the spokes are in place, continue to weave, using whatever material you want. Soft materials, such as baby seagrass or dyed sisal, are useful in separating the spokes at this stage, but you should change to a harder material, like palm inflorescence and sugar palm, at the bottom. Remember, that all spokes should have no more than a two-finger width between them. As you near the bottom, insert the "basket feet" as follows.

Create two "basket feet" to help with balancing the basket by adding two antler slices on separate spokes and insert the spokes into the basket bottom. These should be the second spoke from the outside bottom spoke. If the basket still doesn't sit right, simply move the antler slice until the pouch balances. Weave them in place like you would any new spoke.

Finish the basket by adding some leather strips over the antler and around the ring/rim. Add embellishments such as beads and feathers if desired.

Antler Porkypod

Jill Choate

Materials

- Slightly curved two-point whitetail or muley antler
- #8 round reed
- 1/4" oval/oval reed
- #3 round reed, natural and dyed
- Drill and 1/4" drill bit
- Hot glue gun and masking tape
- Knife, rubber band, clothespin, and weight
- Round nosed pliers and spritzer

Creating the Rim

Select a two-prong whitetail or mule deer antler with a slight curve. Create a curved hoop with the antler by drilling a hole at the butt end of the antler at an angle towards the inside curve. Drill a second hole on this inside curve about 1/2" away from the butt end, so that this new hole meets the first drilled hole. The angle should be so that the #8 round reed can be inserted into this combined hole and feed around the curve of the antler, creating a hoop. The hoop is also taped to the antler at the curve and one antler tip. The actual size of the hoop opening can be determined once the basket base is woven. This basket can come up and inward, or it can flair outward, or dip down at the rim. Because the shaping is up to you, wait to glue and tape the hoop until the base is woven and the basket has started to be woven upwards.

Beginning the Base

This basket utilizes two different methods of weaving, rib construction, and twining to obtain its shape and color. The base of the basket is begun using a twined method. From the 1/4" oval reed, cut four spokes approximately one and half yards long. Mark the center of each spoke. Using a knife, scarf, and indentation in the center of the spokes so that they will fit two-on-two on top of one another. This helps the spokes lay flat when weaving them.

You should now have a cross of four spokes facing you. Begin the weave using a long length of wet #3 round reed that has been bent in half. You will be using a twined weave around two sets of spokes for three rows. Mark the first spoke with a marker of some type. You will refer to this as your start/stop point throughout the entire basket. When three rows are woven, separate the spokes for a total of eight spokes and twine for four more rows. It's now time to add more spokes. One spoke is added to each existing spoke for a total of 16, and the twining is continued, separating these new spokes for four rows. One last round of spokes is added for a total of 24. Weave until the base reaches approximately four inches in diameter.

Photography and copyright by Jill Choate.

Building the Design

After completing the base and before weaving the sides, it's essential to have a weight of some type in the bottom of the basket. Using a three-rod rand for three rows, begin to weave the spokes into an upright position. The basket is flaring out at this point. Add one more weaver at the start/stop point and continue with a four-rod rand for four rows. Soak the upper spokes of the basket with the spritzer so that they will relax.

Cut the "quills" from the dyed reed, assorted colors if desired. These should be about five inches in length. Beginning at the start point, pull one quill around each basket spoke. Wet two lengths of #3 reed and twine over these quills. Continue this quill/twining technique for another three rows of color. Remember to pack each row of color/twine.

Using the needle nose pliers, pinch each spoke so that the basket begins to flare inward. Using a rubber band and a clothespin, hold the spokes together at the top of the basket to assist with this inward flare. Weave three rows of three-rod rand with #3 weavers. Switch to dyed reed and four rod rand for six rows. (You might consider using two strands of two colors for this step.) Switch back to three-rod rand for two rows. You should be about one third done with the design. Now is the time to determine the top antler/hoop shape.

Place the antler/hoop over the basket top. Determine how far up the weaving should go and whether or not the basket top will be level or angled. Also determine how large to make the upper hoop. Cut the #8 to fit that size and glue the hoop ends into the antler. Tape the antler to the hoop as well, both at the curve and at the antler tip. Remove the antler/hoop and weave with natural #3 round reed until the basket is about three quarters finished.

Switching Over

Replace the antler/hoop on the basket. It's time to change gears from twining to rib construction. Decide how you want the rim of the basket to sit, horizontal or at an angle. If you choose to angle the hoop, the antler tends to lie flatter against the sides of the basket.

Using a never-ending five-point lashing (a five-point lashing that continues around the rim of the basket), lash each spoke to the hoop. Begin by dividing the spokes in half so that there is an equal number on each side of the rim/hoop. This lashing may take several different weavers to accomplish this, so just tuck the new ends in and begin with another. Remember to cut off the masking tape when you come to it. Lash around the antler in two or three spots. This will make a sturdy transition from the rim/hoop to the basket.

Once completed with the rim/hoop lashing, weave using filling techniques, from the top of the basket until you achieve a level weave in the basket. You can then switch to a chase weave until you meet the existing weave at the bottom of the basket. To finish, cut the remaining weavers above the lashing, about 1/4" from the top of the rim.

Love of the Painted Gourd

Pam Montgomery

As a self-taught artist for over twenty years, I teach monthly classes at our local hobby store, using gourds as our canvas. I am very involved with the state of Alabama Gourd Society, and actually, that's how I got hooked on gourds. My first attendance at a gourd show was 2003, and as they say, the rest is history. Upon leaving that show, my car was full of as many gourds as it could hold and I never looked back. My love of gourds keeps growing, as does my choice of topics, from Santa's to wild life to Native American designs.

I love incorporating deer antlers into my work. I start by marking and then cutting away the top of the gourd, followed by cleaning out the inside. Then I select the correct antler to fit the gourd. A straight antler or an almost straight one is easiest to work with for this technique.

I drill a small hole on each side of the gourd where the antler will fit through. I start with a small hole and enlarge each side with a Dremel®, as each antler end is different. Work slowly and a little at a time. Be careful to not add too much pressure while you are fitting in the antler or you may crack the gourd. Once it is in place, I remove the antler and proceed with painting the image on the gourd.

My paint of choice is acrylic, which I use both inside and outside the gourd. I also like to add a subtle faux finish, such as sponging or glazing to the outside. This will give the exterior more depth. Then I add the image in pencil. If you are not comfortable with freehand drawing, use a pattern with transfer paper. But try drawing. You will be surprised what you can do with a little practice.

To my image, I add a base coat and then the details. Use several values (variations of light and dark of the same color) to add shadows, gradually building to the depth of the image. For the final details, I use Rembrandt® soft pastels to add those deep wrinkles in the faces, and sometimes to the clothing details as well.

Finally, I mist the painting lightly with a spray fixative made especially for pastels. It is very important to use a very light coat of fixative as the pastels may run. Let it dry five to ten minutes and then varnish the entire gourd exterior. I prefer a brush-on, but you can use a spray, but carefully. Again, overloading with spray can cause drips that will spoil the entire project.

My last step is to insert the antler. If the antler is loose, a small dab of glue at either the butt or point end will help it adhere to the gourd.

Free-Form Antler Sculpture

Betsey Sloan

Materials

- Mule deer antler, about 14 inches
- #5 and #2 smoked round reed
- #2 and #3 assorted dyed round reeds
- Yarns and thin cedar strips
- Vines, honeysuckle, or similar
- Baby seagrass
- SuperGel Glue and scissors
- Drill and bit to fit #5 reed

Working a free-form design on any antler is fun. You just have to think "outside of the box." I like to use mule antlers because they are usually longer and broader than whitetails. First, imaging three or four "basket" sides coming from the points of the antler down to the base. The antler can lie anyway you want, upright on its end, sideways, or with the tips downward with the butt up in the air. However you decide, the weaving has to be able to support the antler. For this weaving, I chose to have the antler lie on its outside curved area.

Using a drill bit to fit the #5 spoke material, start by drilling three holes in a row at the butt end. Drill three corresponding holes in any of the antler tips. Using three lengths of spoke material, feed the reed from butt to tip. Eyeball where these spokes would best create an arch of weaving. The spokes should be no more than two finger widths apart at the outside curve. Additional spokes can be added during the weaving process if needed.

Select another section to have a weaving arch, either over or under the first arch. Drill three holes each side of the antler, feed in lengths of reed. Finally, locate another spot for a third arch of weaving. Again, this new arch can go over and/or under the other two arches. Once you are satisfied with the arches, glue the spokes in place.

I start with the smallest material that I have, but not the yarn or seagrass in case I need to add new spokes. I prefer to use a "hard" material to start so the new spokes have something to slip into that will hold them in position. Weave is a simple over/under. Change materials as desired. A balance between hard and soft materials will keep the arches in position and add a visual balance to the entire piece.

This type of free-formed basket is great for using up odds and ends of weaving material that may be too short or the amount too small for a normal antler basket.

Patience and Flexibility

Leah Reed

Did you ever have a great idea for a project? You know, the one you had at 3 a.m.? You searched for all the pieces that would be perfect for the project. And then, as you began to put them together, realized that nothing was going as planned? Has anyone ever told you that gourd art teaches you patience and flexibility? Well, my project "Lost in the Night" did that for me.

I had this terrific red deer antler and what I thought was the "perfect" gourd. I mounted the gourd to the antler by drilling a hole through the base of the antler and into the bottom the gourd, then fastened them together with a bolt. Agghhhh, what a disappointment! The two looked horrible together. Back to the gourd pile I go, looking for inspiration. I found one that looked like a carrot but felt it might just work.

Now comes the time to be flexible...by letting go of the original idea for the project. The original design just was not going to work with this new carrot shaped gourd, but it did look good with the antler. Well, now patience comes in to play. Both the gourd and the antler went up on a shelf while I waited for inspiration to strike.

I don't recall what the spark was, but once I drew out a design of wolves for the gourd, everything started to come together. I cut the top of the gourd off in a clean cut so it could be reassembled without the cut line showing. The wolves and moon were done with basic relief carving. The gourd surface was burned with a torch (carefully) and polished using 600 through 12,000-grit sandpaper. The moon and insides of the gourd were completed with variegated gold leafing. I attached the gourd top with leather lacing and, lastly, both the gourd and antler were attached to driftwood. I then added the embellishments.

In conclusion, you have probably heard other gourd artists say, "Listen to the gourd and it will tell you what it wants." I have to agree. If you find yourself in this situation, just put the pieces to one side and wait for that "spark" to appear.

Gallery

This free-form antler sculpture (from the author) features a large mule antler, smoked reed as spokes, and dyed and smoked reeds, yarns, cedar, and seagrass as weaving materials. These materials allow the basket and antler to sit straight up.

This carved and woodburned gourd was stained with both acrylics and leather dye. The carved background is painted a light coat of ivory acrylic paint. Small holes were drilled in both the tip and butt end of the antler and in the gourd rim. Waxed linen was used to secure the antler to the gourd rim. Leather lacing hides the attachments. *From Debbie Cain*.

"Warrior Queen" by the author. Coiling with waxed linen around the neck of the gourd with beading at the base. The mask was stained with alcohol inks and acrylics. Two single-point whitetail antlers rise from the top of the gourd. Embellishments such as feather plumes, silver gourd earrings, beads, and bones complete the image.

A large whitetail antler is carefully balanced and secured on this gourd (from Bonnie Cox). A simple feather design is woodburned opposite the antler to balance the overall design.

The graceful lines of a large mule antler is allowed to dominate this gourd. The gourd was grown by her daughter-in-law in Georgia, and the parrot feather is from the farm where she worked. A rim of woven leather completes the gourd. *From Bonnie Cox*.

This artist chose to have the whitetail antler come through the gourd from the inside, which makes for an interesting approach. The gourd was left natural on the outside with a paste wax finish. The antler is secured by the coiled seagrass rim. *From Terry Noxtel*.

Woodburned animals abound around this gourd. The curved whitetail is a pleasing focal point around a woodburned rim, separated from the design by an inlaid strand of beads. *From Bonnie Cox*.

The artist used oxblood and barn grey gourd dye for the base colors on this gourd. She also used fumed pine needles and black waxed linen for the rim to hold in the whitetail shed antler. The finish is polyurethane. *From Cathy Toot*.

In this gourd antler basket, the artist chose Gilders® paste as her colorant. She also coiled a pine needle rim and used the whitetail shed as a base for the artwork. The gourd was finished in polyurethane. *From Cathy Toot*.

"Night Hunter" utilized an undulating gourd rim with an owl design, both carved and woodburned. Tria® alcohol inks were the colorants and a whitetail antler was lashed to the rim with waxed linen. Acrylic spray sealer completed the gourd. *From Linda Barrett*.

"Cabin for Rent" is the whimsical title of this gourd. A whitetail antler has been lashed into a rim of leather lashing. Leather dye has been sealed with acrylic spray. *From Mary Gayle Van Ingen*.

This gourd was created for the 2006 Georgia Gourd Gathering. The artist used acrylics, spray sealer, and a curved whitetail antler, which is resting on the rim. *From Judi Besier*.

"Serenity" seems an apt name for this gourd. Following an application of leather dye and acrylic spray, the artist coiled paper cord with waxed linen, holding in the long graceful whitetail antler. *From Susan Ashley*.

"On Point" was woven with flat and round reed, seagrass, and suede lacing. Embellishments are polished stone, beads, and dried grass. *From Susan Ashley*.

"Follow the Curve" is the title of this gourd. The artist used leather dye finished with spray acrylic. The inside was painted with black acrylic paint, and the coiling is waxed linen over paper cord. A very curved whitetail antler is held in place by the coiling. *From Susan Ashley*.

"Appalachian Antler Heart" utilizes flat and round smoked reed, seagrass, and date palm inflorescence. Embellishments are polished stone and wooden beads. *From Susan Ashley*.

This gourd, called "The Hunt," was created by first staining the gourd with leather dye and then adding the antler with artificial sinew across a pine needle stitched rim. A small Indian spear, pinecones, and a faux pine branch completes the design. *From Betty Banks; photography and copyright by Sarena Kauk.*

"Wild Blackberries" features an inlaid spiral design in InLace® resin. A woodburned basket weave has been added along with a silver concho in the center front and back. The gourd is supported by the points of a shed antler that is not permanently attached. *From Betty Banks; photography and copyright by Sarena Kauk.*

"Searching for a Home" shows the gourd tilted on its side and sitting on a 4-point shed antler, which is fastened onto the gourd with artificial sinew. Leather dye was used as the colorant, and turquoise nuggets and playful clay lizards add the focal interest point. *From Betty Banks; photography and copyright by Sarena Kauk.*

"Around the Campfire" was colored with leather dye and then coiled around the rim with pine needles. A clear acrylic spray finish was topped off with a large mule antler. *From Toni Best; photography and copyright by William Dillberg.*

"Dream Catcher" features a very curved mule antler. The colorants were leather dyes finished with acrylic spray; the rim is coiled pine needles. *From Toni Best; photography and copyright by William Dillberg.*

This "Granny Antler Basket" started with three whitetail antlers rimmed together; three pouch-style baskets were then woven from it. The artist used oval/oval reed plus round reed for the rim, spokes, and weavers. *Photography and copyright by Jill Choate.*

Titled "Lifted from the Earth," a large mule antler sits on top of a leather dyed gourd with a coiled pine needle rim. Finish is spray acrylic. *From Toni Best; photography and copyright by William Dillberg.*

"Elktooth Gourd" has imitation elk teeth, and slices of deer antler used as buttons, tied on with leather thongs. *From Bonnie Gibson; photography and copyright by Everett Gibson.*

This stunning woodburned and carved "Mule Deer" utilizes a set of antlers still attached to a portion of the skull as the stand for the gourd. *From Bonnie Gibson; photography and copyright by Everett Gibson.*

This gourd, "Hunted," has images that are woodburned and then carved, followed by washes of gouache paints. The rim is coiled pine needles using waxed linen, with the curved antler incorporated into the coiling. Light coats of Deft® non-yellowing spray finished the work. *Photography and copyright by Hellen Martin.*

The artist used a mule antler as the stand for the gourd in this project, called "Woven Antler Gourd." Notice that the antler is integral to the design and that one of the antler prongs pierces the antler shell. The antler is also sewn to the bottom of the gourd. A turquoise inlay, scrimshaw, and inlay on the antler button, and a coiled pine needle rim complete the gourd. *From Bonnie Gibson; photography and copyright by Everett Gibson.*

In "Rustic," industrial strength metal bronze paint was used as the colorant. Additional patinas were added for depth. A coiled pine needle rim with waxed linen hold in the curved mule deer antler. Deft non-yellowing spray is the finish. *Photography and copyright by Hellen Martin.*

This gourd, titled "Western Meets Nature," is a woodburned design shaded with acrylic paint plus carving and a minimal amount of black leather dye. She also flowed water-based dye over the design at various places, finishing each flower center with a turquoise rivet. A coiled pine needle rim holds in a curved antler and the finishing material was Deft® non-yellowing spray. *Photography and copyright by Hellen Martin.*

For this gourd, the artist removed the top and side of a gourd and then hung a large mule deer antler over the opening with leather deer hide strips. The gourd is colored with shades of brown acrylic and finished with clear satin spray acrylic finish. *From Nan Noble.*

This gourd sports a large mule antler cleverly hung from the front of an undulating gourd rim. She used acrylic paint sponged onto the gourd, finished with clear satin spray acrylic. The antler is held in place by leather deer hide strips. *From Nan Noble.*

An airbrushed moose in dark brown acrylic paint strides around this gourd. The gourd was additionally colored with sponged acrylic paints and finished in clear satin spray acrylic. The curved mule deer antler rests atop the gourd, held in place by seagrass coiled rim and leather deer hide strips. *From Nan Noble.*

"Deer Reflection" is the title for this gourd. Relief carving and woodburning brought the design to life, inks were the colorant choice, and the antler was incorporated into the pine needle rim. *Photography and copyright by Carolyn Owen.*

"Natural Strength" has woodburned oak leaves on a gourd stained in ink dye. Danish cord and waxed linen was used to weave the antler onto the gourd. *Photography and copyright by Carolyn Owen.*

"Celtic Serenity" has an InLace® resin Inlay beneath the gourd rim and a woodburned and carved knot design. A pine needle rim rests atop the gourd and waxed linen around paper rush securely holds the antler in place to the coiling and to the gourd itself. Embellishments include a turquoise stone suspended between two loops in the coiling. *Photography and copyright by Carolyn Owen.*

Sweetgrass coiling with artificial sinew holds in this curved whitetail antler on this gourd. Leather dye was the colorant and the gourd was finished with spray acrylic. *From Vicki Plienis; photography and copyright by Richard Zimmerman Photography.*

This artist chose a colored cordage as the rim coiling, sewn with artificial sinew. The long single antler was incorporated into the coiling. Leather dye was the colorant and it was finished with spray acrylic. *From Vicki Plienis; photography and copyright by Richard Zimmerman Photography*.

Two antlers intertwined and affixed to the bottom set the base for this gourd. A woodburned design plus pine needle coiling at the sloped rim complete the gourd, which also has a whitetail antler intertwined into the coiling. Alcohol inks were used as colorants and lacquer as the finish. *From Patty Sorenson; photography and copyright by Dave Clements*.

This antler creation includes securely fastening the whitetail into the pine needle coiling. Metal buttons were added for accent while various leather dyes were used for staining the gourd. *Photography and copyright by Gloria Small*.

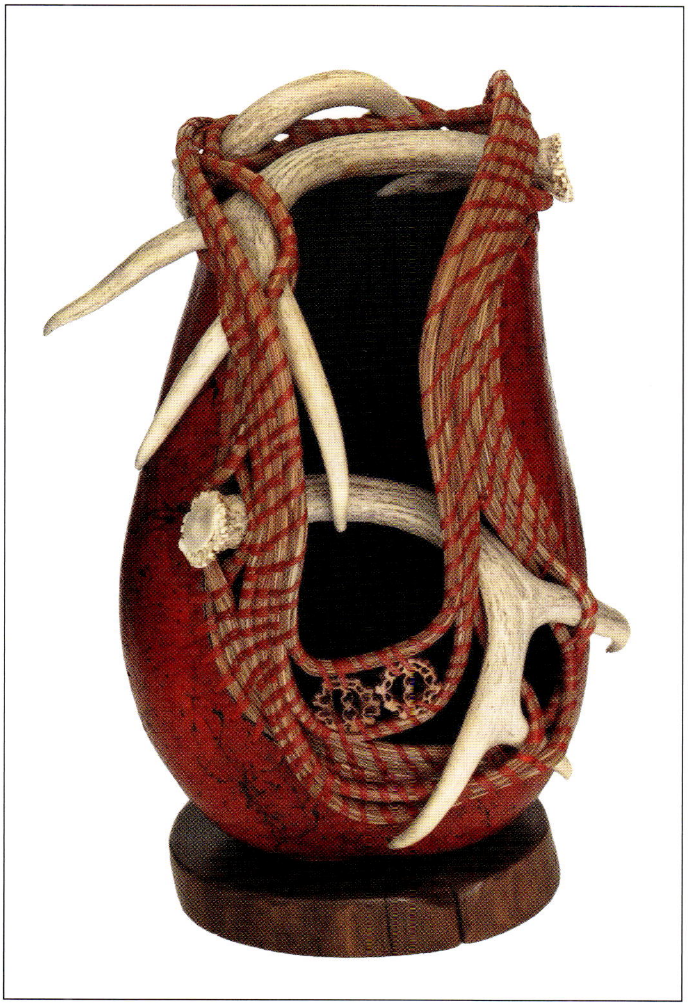

"Scarlet Threesome" has whitetail antlers twined into the pine needle coiling. Alcohol inks were used as the colorant with a lacquer finish. A slice of wood attached at the base helps stabilize the gourd. Several slices of black walnut were intertwined with the coiling for accent. *From Patty Sorenson; photography and copyright by David Clements.*

This carved and woodburned gourd, "Threesome," depicts three deer images. Alcohol inks were the colorant choice with a lacquer finish. The rim treatment is an intricate coiling of pine needles, which securely hold in the antler. *From Patty Sorenson; photography and copyright by David Clements.*

This natural gourc with pine needle coiling is titled "Field and Forest." The natural shed antler, which the artist found, is incorporated into the coiling and a betel nut slice is used as an embellishment. *From Ann Wisener; photography and copyright by David Gill-Lawton.*

"Eric the Red" was inspired by a design from a tattoo book. After woodburning the design, the artist used leather dye as the colorant. He then carved a Celtic design and sealed the entire piece with tung oil. The antler was secured to the gourd using a bamboo skewer, a metal rod, and a little glue. *From Jim Dixon; photography and copyright by Herb Jones, II.*

The end of the antler button has an inlaid Indianhead penny.

Titled "Amanda" for the artist's granddaughter, this flower canteen gourd was woodburned and then stained with leather dye. Parts of the design were carved away and then tung oil applied. Some of the flower designs contain antique buttons. The antler is secured using bamboo skewers through both the gourd base and antler. *From Jim Dixon; photography and copyright by Herb Jones, II.*

This gourd, titled "The Emperor," uses both woodburning and carving to create an intricate basket weave and rope design with inlaid leather lacing. Leather dyes were the colorant and tung oil provided the finish. The antler is secured to the gourd base by bamboo skewers and glue. *From Jim Dixon; photography and copyright by Herb Jones, II.*

"Indian Summer" uses a canteen gourd set on its side, resting and attached to a deer antler. An intricate basket weave design was created by woodburning and leather dye with a tung oil finish completed the gourd. A large piece of turquoise was set in the center as the focal point with the antler secured by bamboo skewers to the gourd base. *From Jim Dixon; photography and copyright by Herb Jones, II.*

Titled "Antler Afoot," this basket uses a multi-pronged whitetail as the basket base, with the prongs pointed towards the table. Using dyed and natural reeds, leather, and other natural materials, the basket rises from the antler. *From Nancy Jones.*

"Forested Antler" utilizes round reeds and natural materials. An antler is woven into the basket rim as the focal point. *From Nancy Jones.*

"Picking Berries" uses only one colorant, black leather dye for the interior. Tung oil is the finish on the natural exterior. Bald cypress branches, leather ties, jade pendants, and brass beads complete the gourd. The antler is secured to the basket rim with artificial sinew through holes in the antler and the gourd. *From Jim Dixon; photography and copyright by Herb Jones, II.*

Adding antlers to be used as the handle, this artist created this basket from flat natural and dyed reeds. The antler is secured to the basket rim by strips of flat reed woven over the butt and tip ends of the antler. *From Joyce Storto*.

The artist created this whimsical gourd with antler from a canteen gourd dyed with leather dyes. Feathers abound at the gourd opening along with hickory nut shells. *From Jaynie Barnes*.

Natural mushrooms were used on this gourd. The artist started with a canteen gourd dyed with leather dye, and then added the deer antler upside down at the gourd rim, tied on with waxed linen. Some glue was used to attach the side mushrooms, and the finish on the gourd was spray acrylic. *From Jaynie Barnes*.

This artist added two small, matched whitetail antlers to the rim of this basket once it was woven. She used natural and dyed flat reeds for the weaving pattern. The antlers are secured in position by strips of flat reed woven over the antler and through the basket rim. *From Joyce Storto*.

This gourd, titled "Red and Black," was stained with Old English® furniture polish followed by a spray acrylic. A pine needle rim was topped by a whitetail antler, which was drilled and lashed into position with waxed linen. Black and red beads lend the gourd its name. *From Arlene Zaloznik*.

Using a pear-shaped gourd as the base, the artist first stained the outside with ink dye and the inside with leather dye. After coiling a pine needle rim, she added two straight whitetail antlers to the gourd rim and added beads, hickory nuts, and a mini gourd between the antler tips. A large natural mushroom was glued on as the gourd base. *From Jaynie Barnes*.

"Elegant" is the name of this gourd. The artist used Old English® furniture polish as the stain and a clear acrylic sealer. Philodendron sheaths with an antler button (slice) adorn the rim along with beads. *From Arlene Zaloznik*.

Titled "Peace," this gourd was first stained with gourd dye and then sprayed with clear acrylic. Pine needles were coiled along the severely cut gourd rim. The antler was attached to the rim through small holes made in the antler at three points. *From Arlene Zaloznik*.

This undulating basket/bowl is titled "Royalty." Round reed spokes were glued into holes around the antler and dyed reed and seagrass were woven to create the basket sides. *From Arlene Zaloznik*.

"Bird in Flight" shows the good use of a slightly deformed antler. The artist often works such antlers into her basket weaving. Here she has allowed the antler to dictate the position of the finished artwork. *From Karen Zane*.

Using a very curved whitetail antler as the base, the artist wove this basket/bowl titled "Fall" with dyed round reed and dyed seagrass. Holes were drilled along the antler sides, which held the round reed spokes in position. *From Arlene Zaloznik*.

"The Tunnel" creates the impression of entering into the basket as it is resting on the tines of the mule deer antler. Round reeds create both spokes and weavers. The artist allows each antler to inspire her baskets and believes there is no wrong or right way to create with antlers. *From Karen Zane*.

This gourd was lightly stained with gourd dye inside and out before a finish of acrylic spray was added. The large whitetail antler is attached at two points to the rim, allowing the butt end to rest on the table. *From Jill Nelson*.

This gourd, with a leather rim, shows the effective use of turning an antler on its side to achieve the focal point in the piece. After using colorants of both gourd and leather dye, the antler was securely attached to the gourd at both the butt and point end, and the addition of feathers completes the work. *From Jill Nelson*.

"The Warrior," inspired by Native American traditions, uses whitetail antlers, turkey and pheasant feathers, raccoon and coyote fur, horse hair, glass beads, and turquoise. The antlers are bolted to the gourd and a special attachment created with wood and dowels holds the mask on a stand. *From Jill Nelson*.

This artist's gourds often include Native American designs painted with acrylics and highlighted with oil pastels for depth. She uses a fixative over her artwork and then a coat of varnish. The deer antler holes were drilled first, then the antler removed and the design applied and finished. The antler was then inserted into the holes and secured with a small amount of glue. *From Pam Montgomery*.

"Deer Tracks" uses one of the antler points to balance the gourd. Gourd dye was used as the colorant along with leather trim around the rim. Clear satin acrylic spray was the finish, with a turquoise stone set into the antler as an accent. *From Gloria Crane*.

This "Coiled Antler Bowl" has a deep coiled rim of pine needles with artificial sinew. The mule deer antler is secured in the coiling and a Red Creek Jasper stone was inlaid in the antler at the fork. *From Gloria Crane*.

This "Antler Jewel Box" was stained with gourd dye plus metal patina and copper leafing. The finish is clear satin acrylic. Deer skin lines the inside of the gourd while a curved deer antler, secured with leather lacing, runs over the gourd top. *From Gloria Crane*.

This large basket was designed and woven on a mule antler. Round and flat reeds were used for the spokes and weavers, and multiple beads create the unique focal point. *From Fonda Haddad*.

A natural colored gourd holds a forked mule antler in this design. The antler is secured at two points — the gourd rim and butt end. Numerous beads and turquoise were added to embellish the gourd. *From Fonda Haddad.*

This "Horsehair Antler Basket" was first woodburned and then colored with wood stain. The curved whitetail antler was secured with waxed linen and assorted beads and stones were added as embellishments. *Photography and copyright by Shelley Harris-Janac.*

This gourd, "Four Feathers," was stained with leather dye. A whitetail antler was allowed to curve over the gourd rim and was tied into place. Feathers, beads, and turquoise nuggets were added as accent pieces. *Photography and copyright by Shelley Harris-Janac.*

"Pine Cones" sports just that — multiple cones and pine needles around the rim of this gourd. Acrylic paint was used as the colorant and the whitetail antler is woven into the coiled rim. *Photography and copyright by Shelley Harris-Janac.*

"Four Stars," a key feature of Native American spiritual design, was incorporated into this design. Leather dye was used as the colorant, and the whitetail deer antler was secured to the gourd with lacing. *Photography and copyright by Shelley Harris-Janac.*

Aptly named "Squirrel Gourd," this gourd was chewed by the varmint during the gourd drying process. Not one to waste a gourd, the artist stained it with leather dye and used acrylic paint on the inside, acrylic sealer on the outside, and added a small antler to the top as the accent. *Photography and copyright by Shelley Harris-Janac.*

"Turquoise Antler" was coated with paste wax plus an acrylic paint on the interior. Turquoise nuggets and beads were added along with a small whitetail antler, held to the rim by lashing over the antler to the gourd rim. *Photography and copyright by Shelley Harris-Janac.*

Titled "Turquoise Lights," this gourd was finished with a copper metal coating, using several patinas to create the finish, which took several days to achieve. Coiled pine needle hold a small moose antler in place. Turquoise beads were strung across the antler for embellishment. *From Judy Richie.*

"Accents in Bronze" has as its focal point a slab of Jasper inlaid in the small moose antler. The colorants were bronze metal and several patinas over a woodburned design. The coiled pine needle rim holds the antler securely in place. *From Judy Richie.*

"Pueblo Story" is a gourd rich in the Pueblo culture with hand-painted designs. The rim is coiled pine needles encapsulating a curved mule deer antler. *From Judy Richie.*

This very large gourd, titled "Copper Rack," uses copper metal coating and several coats of patina. The coiled pine needle rim and metal ribbon holds in a mule deer antler. *From Judy Richie.*

The author's "Zuni Ceremonial Bowls" uses a combination of woodburning, InLace®, Tria® inks, and acrylics for the design and colorant elements. The rims are coiled seagrass with handmade fetishes plus bones, beads, and leather as embellishments All three gourd bowls sit atop and are attached to an elk antler.

Biographies

Susan Ashley
Email: susanm@txweaver.com
Web: www.txweaver.com

Susan took her first basket weaving class in 1983 and fell in love with the craft. Growing up in West Virginia, the Appalachian egg baskets just seemed to come naturally to her. By the following year she was weaving and selling her work at juried shows. For many years she attended the annual Michigan Basket Weaving Conventions, where she studied with many different teachers and explored new techniques and materials. It was on one of those trips a few years ago that she took a class with Marla Helton and became fascinated with weaving on gourds.

Susan says that weaving baskets and weaving on gourds fills her need to make something that can be "physically touched and felt" at the end of the day since she can't really do that in her 'day job' as a systems analyst. She is drawn to things that are symmetrical and her designs are usually very simple, incorporating natural elements. Awards include ribbons from the Texas State Fair and Lone Star Gourd Festival.

Marianne Barnes
Email: maribasket@charter.net
Web: www.maribasket.com

Marianne, also know as Maribasket, graduated from Winthrop University with a double major in art and education. She has been working with gourds for the last twenty years. She has been a basket maker for twenty-one years and an art teacher in the public schools for thirty-eight years. Barnes retired from teaching in 2008.

The combination of baskets and gourds is the art form with which Marianne most loves to work, texture being the core of her art. She works with natural and textural materials such as cedar and birch bark, seagrass, mohair and knobby yarns, pods, philodendron sheath, and leather. She also incorporates basketry by weaving on top of the gourd with an undulating tapestry weave that resembles a landscape. Recently Marianne has incorporated weaving with round reed and adding zigzags, arrows, spirals, and darts in different colors. She has also been adding different techniques to enhance the gourd itself, such as woodburning, painting, and carving. Many times a dream catcher or a tapestry weave on the front of the gourd is added.

Judy Briscoe
Email: basketwman@comcast.net

Although working in fiber for many years, Judy has been a basket maker since 1990. She specializes in baskets and gourds inspired by nature and uses both native and other natural fibers in her work. Her favorite materials include inflorescence, philodendron sheaths, driftwood, cypress knees, antlers, and barks from western red and yellow cedar, elm birch, and red pine.

Two of her baskets are on permanent display as part of the "Heritage of Baskets Project" at the Michigan State University Museum. In 1996, Briscoe was chosen by the Smithsonian Institute's Center for Folk Life Programs and Studies to be included in the "Southern Crossroads Marketplace," a venue for the sale of southern arts and crafts at the 1996 Olympic Games held in Atlanta, Georgia. She teaches out of her home studio and at conventions, basket shops, guilds, and art centers around the country.

Jill Choate
Email: jill@jchoatebasketry.com
Web: www.jchoatebasketry.com

Jill is a nationally renowned fiber artist and instructor, known for her ability to "talk to the antler" and weave a vessel around it. Her pioneering efforts incorporating antler into basketry have brought her to the forefront in this venue of contemporary vessel construction. Her efforts in basketry are as well known as her stories about life in the Alaskan bush.

Now based in Van Buren, Missouri, she lived in remote Alaska for twenty-five years with her family and sled dog team approximately 150 miles from Anchorage. The Choate homestead was accessible by snow machine in the winter and four-wheel in the summer, and was equipped with all the modern "bush" conveniences...like a hand-pump and an outhouse. Living in the shadow of Denali with a salmon stream in the front yard plus moose and bears for neighbors makes for great reminiscent tales about the Last Frontier.

Jill believes that each creation is woven with a boreal forest in mind. The antler's natural curve, coloration, and texture are taken into consideration, and ignite the potential for a creative vessel to emerge. The branch, the smell of willow, brightly colored leaves, the wind passing through the leaves, snow sifting from the sky... Using the inspirations of these simple gifts, the result is a vessel of woven nature.

Bonnie Gibson
Email: bonniegibson@qwest.net
Web: www.arizonagourds.com

Bonnie is an accomplished artist in many forms of three-dimensional arts and fine crafts, and she has gained recognition and awards for her work in gourds, wood carving, scrimshaw, and scale miniatures.

A self-taught artisan, Bonnie prefers this term to "artist" because to her it denotes a person who works with items that can be functional as well as aesthetically pleasing. Since she was a child, working with her hands and creating things captivated her. Bonnie took art classes in high school, but always felt more comfortable with three-dimensional creations instead of traditional paintings and drawings.

She discovered gourds quite by accident a few years ago, and immediately fell in love with them. Working with gourds allows Bonnie to utilize the technical skills that she developed over the years working with other media. She has lived in Arizona for almost thirty years, so interpreting some of the Native cultures and Southwestern designs are among her favorite subjects. Bonnie utilizes many tools and techniques to reach a finished design, including carving, pyrography, painting, and inlays of stone. Certain design elements — carved sand ripples and basketry effects — have become her signature patterns.

Bonnie has contributed to many gourd crafting books and has written numerous articles on gourd crafting. Her first gourd-crafting book, *Gourds: Southwest Gourd Techniques & Projects from Simple to Sophisticated*, was published in 2006. One of Bonnie's primary goals is to help lift gourds out of the realm of "crafts" and into greater acceptance as fine art. To that end, she enjoys manipulating gourds in new ways, inviting the viewer to interpret them as something more while retaining their natural essence.

Fonda Haddad
Email: fonda-a-art@earthlink.net

Fonda is a professional fiber artist. She has a Master's degree in Visual Arts Education and is a retired teacher. Fonda has had her own original clothing shop, "The Fig Leaf," where she created and sold batik and silk painted clothing. Another thread that has run through her life is basketry. She has created, taught, and sold baskets at the J.C. Campbell Folk School in North Carolina and the Blue Ridge Mountain Arts Association in Georgia. Fonda teaches at many gourd conferences, including Florida, Georgia, North Carolina, Mississippi, Tennessee, and the Florida Tropical Weavers Guild Conferences. Her gourd work is for sale at the Mountain Morning Gallery in Andrews, North Carolina; the Shooting Creek Basketry Guild; the Far Western NC Gourd Patch; and at the Tapoka Lodge near Fontana Dam, North Carolina.

Bob Hosea
Email: Tre9036@sbcglobal.net

Bob is a full-time biologist with the State of California. He has been weaving baskets for over twenty years and occasionally teaches classes at The Caning Shop in Berkeley, California. His baskets are held in private collections around the United States, Europe, and Australia. Bob weaves both utilitarian styles of baskets as well as purely sculptural art pieces. Most of Bob's baskets tend to be made using natural weaving materials that he has collected and prepared himself, such as seaweed, reed, pink palm, honeysuckle vine, seagrass, driftwood, and deer antlers. To embellish his baskets, Bob may use feathers, beads, or other odd items of interest. Using procion dyes and self-collected natural materials, Bob prefers to dye his own weaving supplies.

Over the last five or six years, Bob has been exploring recycled materials in his baskets to make them more eco-friendly. He uses plastic bags from trips to the grocery store as well as plastic strapping, reclaimed wire, and cordage from hay bales. Many of his sculptural baskets have a deep personal meaning, like the seaweed basket he made while his wife was pregnant with their daughter, entitled "Mother Earth." Primarily he enjoys basket weaving because it is relaxing as well as creative.

Pam Montgomery
Email: pamsgourds@bellsouth.net
Web: www.trinitymtngourds.com

Pam has been painting and teaching for over twenty years. She is a self-taught artist, teaching monthly classes at her local craft store in Decatur, Alabama, with gourds as her canvas. Each fall she also teaches 150 fifth and sixth graders a gourd project at one of the local schools. She is very involved with the Alabama Gourd Society. In fact, that's how she got hooked on gourds. She attended her first gourd show in 2003 and, as they say, the rest is history. Pam says her love of gourds keeps growing and her favorite topics are Santa's, wildlife, and Native American designs. Her husband, David, is her biggest supporter and each year they combine a vacation and their love of travel with the Arizona Wuertz Gourd Farm Festival. She also frequents the Alabama Gourd Show.

Nan Noble
Email: nan@nansgourds.com
Web: www.nansgourds.com

Nan was a fiber artist working with teddy bear designs from 1984 to 1995, when her husband's failing health caused her to switch directions. She first raised gourds in 1996 and, of course, was so amazed at their diversity that she "just had to do something with them." Her first big gourd show was the Running of the Gourds at the Pinal Country Fairgrounds in Arizona in 2004 and she became even more intrigued in the magic of gourds. She teaches classes in her RV Park in Arizona during the winter months and sells her work at gift shops in Mesa, Arizona, and Wellington, Nevada.

Leah Reed
Email: leah@bluewhalearts.com
Web: www.bluewhalearts.com

Leah is a national award-winning gourd and egg artist. Although initially self-taught — she started working with gourds and eggs in 2000 — she has studied with nationally renowned artists. Her techniques include carving, dyes, inks, woodburning, oil pencils, coiling, and rim ornamentation.

Leah started Blue Whale Arts, LLC in 2004, selling her fine art at local shows, galleries, and through her website. Because she found it very frustrating that the supplies needed for working with these mediums were not easily found locally, she expanded the business to include selling a wide variety of art supplies and tools. She also teaches gourd classes in her home and at gourd festivals and private venues.

Laraine Short
Email: agourdpainter@aol.com

Laraine is a member of the National Society of Decorative Painters and past president of the local chapter, Decorative Artist of Jacksonville, Florida. She is president of the Florida Gourd Society and the Northeast Florida Gourd Patch. Laraine is also a member of the South Carolina and Georgia Gourd Societies.

She teaches decorative painting and gourd art at her studio as well as at local and state painting chapters, art shows, art stores, local and state gourd patches, and state and national conventions.

Laraine likes to create original designs, some of which have been published in craft magazines and books. She likes traveling to different gourd venues and has been very honored to receive blue ribbons at the Florida, Georgia, North Carolina, Alabama, and Ohio shows. Her last accomplishment was to receive "Best in Show" at the Florida, Georgia, and North Carolina shows.

JoAnn Vanderheite
Email: poojo2@roadrunner.com

JoAnn is a retired art educator who has worked with gourds as an art form for close to ten years. A familiarity with art processes, techniques, and materials, plus her interest in various cultural and historical arts, has helped to shape her collective art experience. With an art background coupled with a creative drive to explore the possibilities of the natural gourd, she has produced many award winning pieces.

JoAnn regularly exhibits her work at juried art shows at a local gallery and at the Ohio Gourd Show. She is a member of the American Gourd Society and the Ohio Gourd Society. She lives in New York State where she is able to grow many of her own gourds, thus carrying her process from seed to finished art.

Mary Gayle VanIngen
Email: allagourd@comcast.net

Mary Gayle fell in love — *with a GOURD* — while on vacation in Albuquerque, and that love has taken her to unexpected places far from her roots in Eastern Montana and her work life in Oregon. She was obsessed. She wanted to know everything there was to know about this intriguing brown vegetable. Mary Gayle joined the American Gourd Society as well as several state chapters. She got into gardening, of all things, and learned to listen to those teeny gourd voices telling her what they wanted to be. She went through a huge learning curve as she sought out every tool, tip, and technique available. She learned about fibers and how to use materials. She also joined a local basketry guild and served on a board of a local committee promoting the arts. She developed a signature style and then began to teach.

Mary Gayle's gourd art has been featured in various galleries, on promotional brochures and calendars, and many of her pieces are in private collections. She is available as an instructor and can be found at various gourd shows across the country.

Karen Zane
Email: kzanebskt@aol.com
Web: www.karensbaskets.com

Karen started making baskets in 1991. She has had the good fortune to study under Judy Wilson for four years where she learned Judy's "36 Rib Buttocks Baskets." With her consent, Karen started teaching at the AMB Convention in 1999 and has never looked back. She has designed several of her own baskets and has had three booklets self-published. They are *The 36 Rib Basket*, *The Basic Rib Basket Book*, and *Antlers Made Easy*. Rib construction is her focus in basketry, and she enjoys the free form of antler baskets as well as the creativity they bring to the classroom.

Supplies Directory

While this is not a complete list of gourd art and basketry suppliers, it is a list of those I use most often and that carry the products, tools and equipment found throughout this book. The photographer for most of the images in this book is also listed.

Blue Whale Arts
Leah Reed
www.bluewhalearts.com

General gourd art supplies and tools including woodburners, burs, smaller-sized respirators, sanding sticks, Micro-mesh 3" sanding disc kits (600-12000 grit), Ostrich eggs (and all the supplies required to work with them), embellishments, inks and paints, red deer antlers, birch bark, and inlays from around the world. It also has hand-dyed feathers and pine needles.

The Caning Shop
Jim Widess
www.thecaningshop.com

Provides supplies, tools, and books for gourd crafters, basket makers, and chair caning.

Eidnes Furs, Inc.
www.eidnesfurs.com

Features all types of antlers, fur, leather, bones, animal parts, antler buttons, and much more.

Ghost Creek Gourds
Dickie and Linda Martin
www.ghostcreekgourds.com

Provides a good assortment of gourd shapes and sizes. Shipping is available. Indicate diameter and thickness desired.

Bonnie Gibson
www.arizonagourds.com

Foredom carvers, numerous bits, woodcarving and woodburning tools, embellishments, beads, drum and musical supplies, gourd books, tutorials, monthly newsletter, and much more.

Gratiot Lake Basketry
Carole Kaeding
www.gratiotlakebasketry.com

Hand pounded black ash, dyed seagrass, natural materials including birch bark, elm bark, pine needles, cane, and other basket weaving supplies.

Suzanne Moore's North Carolina Basket Works, www.ncbasketworks.com

Provides basketry and chair caning supplies, handles, seagrass and other weaving materials, Nantucket Lightship molds, and accessories.

Primitive Originals
Kathy and Bob James
www.primitiveoriginals.com

Features an extensive line of gourd craft supplies, tools, and equipment, including hand-held KumiHimo KumiLoom, books, Washi paper, beads, and bones.

Royalwood, Ltd., www.royalwoodltd.com

Provides basket supplies, Irish waxed linen, seagrass natural and dyed, and additional basket materials.

Betsey Sloan
www.thepodlady.com

Provides pods and dried botanicals from around the world, deer antlers, dried sea urchins and starfish, dyed reed, basketry kits for gourds, and more for use as gourd rim embellishments or general arts and crafting.

Timeless Memories
Jennifer VanAllen
www.jvanallen.photobiz.com

Provides photography, business cards, posters, custom cards, and bookmarks.

Turtle Feathers
David and Ronna Wuttke
www.turtlefeathers.net

InLace family of products, Proxxon®, Microlux®, and Optima® tools, bits, rotary sander and sanding discs, fade resistant leather dyes, QuikWood®, graphite paper, Angelus® acrylic leather paint, and many more art supplies for the gourd artist.

Bibliography

Peterson, David. *Racks*. Santa Barbara, California: Capra Press, 1991.